Flow Cytometry Lab Protocols

Protocols from the International Cytometry Workshops

www.cytometryworkshops.com

First Edition

Awtar Krishan

Pathology Department
University of Miami Miller School of Medicine

PREFACE

Laser flow cytometry has become a major and important tool for rapid analysis of cell cycle, proliferation and phenotypic marker expression in cells. As an analytical tool, it is unique for rapid and multiparametric analysis of a large number of cells. Thus one can monitor expression of multiple markers and if needed selected sub-populations can be sorted on the basis of their marker expression for growth in vitro or in vivo. For diagnosis and classification of hematopoietic malignancies and enumeration of stem cells in bone marrow transplantation, laser flow cytometry has become the standard method of choice.

Since 2001, a dedicated group of researchers from the USA, UK, Australia, Singapore and Canada have joined their colleagues in India, Turkey and Malaysia to teach the latest applications of laser flow cytometry in biomedical research (www.cytometryworkshops.com). These workshops showcase the latest in instrumentation, software and applications in basic and clinical research. Previous workshops have concentrated on applications of flow cytometry in cell and molecular biology, pharmacology, diagnostics, cancer biology, infectious and parasitic diseases and stem cell research and tissue regeneration. In these weeklong workshops, our faculty present lectures, give tutorials and conduct wet labs for demonstrating various flow cytometric techniques.

Flow cytometry involves use of fluorochromes, antibodies and relatively sophisticated electronics. Thus the need for time tested protocols which can guide a novice is paramount. There are several books in the market which list the protocols and recipes for flow cytometric analysis. Although these are great resources, the cost (hundreds of dollars) makes them unavailable to the students in developing countries. Our faculty agreed that a protocol manual which could list most of the commonly used protocols but was within the means of students from the developing countries was much needed. To offset this deficiency, they contributed protocols they use in their labs and teach in the workshop wet labs for use of our students. We have collected 20 of these protocols in the present format. We intend to provide this protocol book to our workshop students as part of their registration package where as others can purchase it for a nominal cost. Any funds collected from sale of this book will be used to support the workshops.

The current edition has protocols written by some of the pioneers and well-known experts in flow cytometry. They have used their data, knowledge and expertise to describe protocols for flow cytometric analysis of cells. It is sincerely hoped that the resulting effort will serve the intended purpose and will help our students to use the standard and time tested protocols developed by our faculty.

We are thankful to the International Society for Advancement of Cytology (ISAC), International Society for Clinical Cytometry (ICCS) and International Union against Cancer (UICC) who have consistently supported our teaching efforts in developing countries. We are also thankful to Mr. Ronald Hamelik who painstakingly edited and formatted the text.

Awtar Krishan, USA.

TABLE OF CONTENTS

Basics

Cell Cycle, Proliferation and Apoptosis

Functional Assays

Stem Cells

Miscellaneous

CONTRIBUTORS

Badrinath, Y

ybadrinath@rediffmail.com
Hematopathology Laboratory, Department of Pathology,
Tata Memorial Hospital, Mumbai, India

Cabana, Raquel

rcabana@blueoceanbio.com
Blue Ocean Biomedical 1 Southwest 129th Avenue, Suite 201
Pembroke Pines, FL, 33027 USA

Cram, L. Scott

scottcram42@gmail.com
Los Alamos National Laboratory, Los Alamos, NM, 87545 USA

Dikshit, Madhu

madhu_dikshit@cdri.res.in
Cardiovascular Pharmacology Unit, Central Drug Research Institute,
Lucknow, India

Gujral, Sumeet

s_gujral@hotmail.com
Hematopathology Laboratory, Department of Pathology,
Tata Memorial Hospital, Mumbai, India

Gupta, Maya

Maya_rk@rediffmail.com
National Institute of Immunohaematology, ICMR,
13th Floor, KEM Hospital,
Parel, Mumbai 400 012, India

Hamelik, Ronald M.

rhamelik@med.miami.edu
Pathology Department R-71, University of Miami Miller School of
Medicine,
PO Box 016960, Miami, FL, 33101 USA

Jain, Navya

navya@ncbs.res.in

National Centre for Biological Sciences, Tata Institute of Fundamental Research, GKVK campus, Bellary Road, Bangalore 560 065, India.

Kalappurakkal, Joseph M.

joseph@ncbs.res.in

National Centre for Biological Sciences, Tata Institute of Fundamental Research, GKVK campus, Bellary Road, Bangalore 560 065, India.

Keeney, Michael

Mike.Keeney@LHSC.ON.CA

Department of Hematology, D1-208, Victoria Hospital, 800 Commissioners Road East, London, Ontario N6A 4G5 Canada.

Kode, Jyoti

jkode@actrec.gov.in

Chiplunkar Lab, Advanced Centre for Treatment, Research & Education in Cancer, Tata Memorial Centre, Kharghar, Navi Mumbai 410 210, India.

Krishan, Awtar

akrishan@med.miami.edu

Pathology Department R-71, University of Miami Miller School of Medicine, PO Box 016960, Miami, FL, 33101 USA.

Krishnamurthy, H

krishna@ncbs.res.in

National Centre for Biological Sciences, Tata Institute of Fundamental Research, GKVK campus, Bellary Road, Bangalore 560 065, India.

Kumar, Ashok

Ashok.san2010@gmail.com

Hematopathology Laboratory, Department of Pathology, Tata Memorial Hospital, Mumbai, India

Kumar, Sachin

sach_bio@yahoo.co.in

Cardiovascular Pharmacology Unit, Central Drug Research Institute, Lucknow, India

Madkaikar, Manisha
madkaikarm@icmr.org.in
National Institute of Immunohaematology, ICMR,
13th Floor, KEM Hospital,
Parel, Mumbai 400 012, India

Ormerod, Michael G.
m.g.ormerod@btinternet.com
1 Furze Place, Furze Hill, Redhill,
RH1 1ER, UK

Sanders, Claire
csanders@lanl.gov
Los Alamos National Labs. Los Alamos, New Mexico, NM, 87545 USA

Sehgal, A.
kunalsehgal@gmail.com
Hematopathology Laboratory, Department of Pathology,
Tata Memorial Hospital, Mumbai, India

Srinag, B S
srinag@ncbs.res.in
National Centre for Biological Sciences,
Tata Institute of Fundamental Research, GKVK campus, Bellary Road,
Bangalore 560 065, India.

Subramanian, P G
pgs_mani@yahoo.com
Hematopathology Laboratory, Department of Pathology,
Tata Memorial Hospital, Mumbai, India

Sutherland, Robert
rob.sutherland@utoronto.ca
Laboratory Medicine Program, Toronto General Hospital, Ontario, Canada

Tanavde, Vivek
vivek@bii.a-star.edu.sg
Genome and Gene Expression Analysis Division, Bioinformatics Institute,
Agency for Science Technology and Research (A*STAR), 30 Biopolis
Street, #07-01, Matrix Singapore 138671.

Telford, William

telfordw@mail.nih.gov

Flow Cytometry Core Laboratory, Experimental Transplantation and Immunology Branch, National Cancer Institute, National Institutes of Health, Bethesda, MD 20892 USA.

1. Basics of Flow Cytometry

H Krishnamurthy
krishna@ncbs.res.in
Scott L. Cram
scottcram42@gmail.com

Introduction
Flow cytometry and cell sorting are unique techniques that permit the identification, analysis and purification of individual cells based on the expression of specific markers.

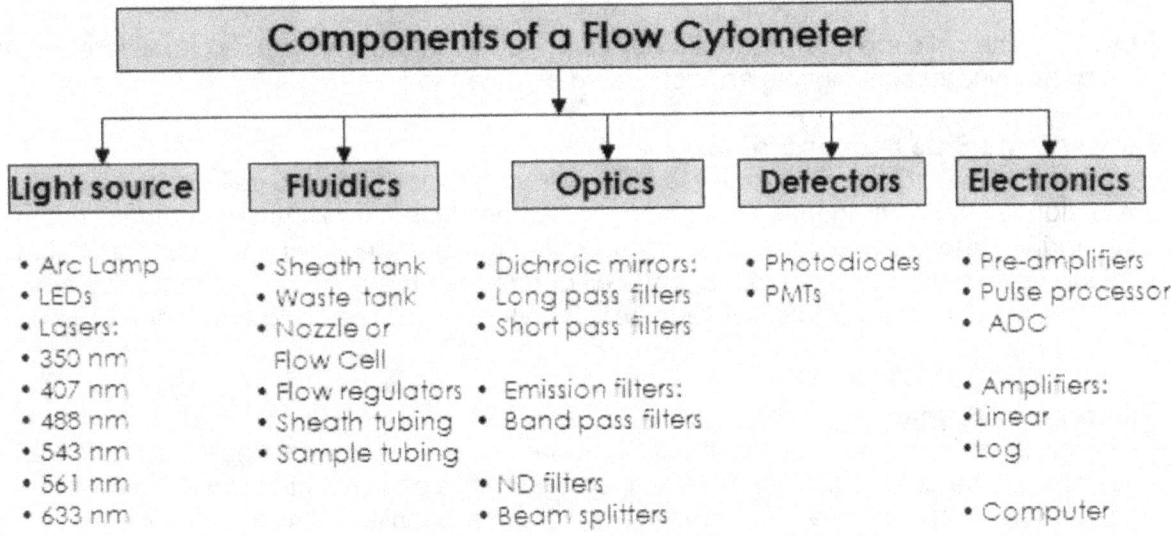

How does a flow cytometer work?
- For flow cytometric analysis, cells (live or fixed) must be in a single cell suspension.
- Cells pass in a single-file through a laser beam (Figure 1).
- Cells scatter some of the laser light and also emit fluorescence from laser excitation of the fluorochrome used to label the cell.
- The cytometer typically measures several parameters simultaneously for each cell.
- The laser beam is focused to the size of a few cell diameters.
- The electronics quantify the scattered light and fluorescent emissions.
- A computer records data for thousands of cells per sample and displays the data graphically.

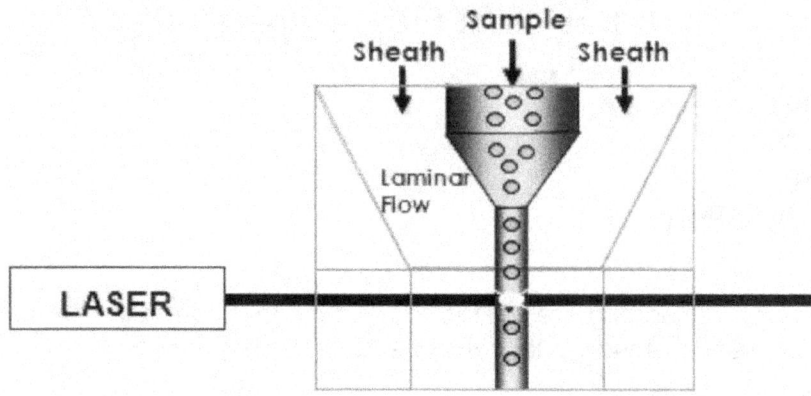

Figure 1. The cells are interrogated by the laser beam in the quartz flow cell where scatter and fluorescent signals are generated simultaneously.

Forward and Side Scatter Signals

Forward angle light scatter intensity is approximately proportional to cell diameter or size, while side scatter (orthogonal, right angle, 90°) intensity is approximately proportional to granularity of the cell and surface complexity. Light scatter alone is often quite useful to exclude dead cells, cell aggregates, and cell debris from the data. Forward versus side scatter is used to discriminate between lymphocytes, monocytes and granulocytes in a blood sample.

Fluorescent Signals

Fluorochromes are used to label specific components (e.g., DNA, antigens, proteins, enzymes) of the cells. Fluorochrome tagged antibodies are used to monitor expression and antigen density of specific surface receptors that enable one to detect discrete subpopulations in a heterogeneous population such as that of bone marrow or a tumor. Intracellular components such as nuclear DNA, RNA or protein content or enzyme activity can also be reported by use of fluorescent probes. Some of the common uses of flow cytometry are for quantitation of DNA content for determination of cell cycle phase distribution and aneuploidy, identification of proliferating cells after incorporation of bromodeoxyuridine, determination of specific nucleotide sequences in DNA or mRNA, filamentous actin, and any structure for which an antibody is available. Flow cytometry can also monitor rapid changes in intracellular calcium flux, membrane potential, pH, or transport or efflux of fluorescent dyes in multidrug resistant and stem cells.

Data Plots

Single parameter flow cytometry data is presented as frequency distribution histograms showing signal intensity on the x-axis and counts on the y-axis. Whereas, dual parameter data is generally presented as dot plots or contour plots showing signal intensity of two different parameters on the x and y axis. The electronic pulse generated by a cell passing through the laser beam is digitalized. However, there will be subtle differences in the signal intensity of the cells of the same population and hence practically the distribution will appear as a frequency distribution and there will be peaks representing the small and big cell populations. The x-axis shows intensity of the pulse and y-axis the number of cells per channel.

Linear and Log Scales
If the range of fluorescence intensity in a population is within a decade, for example the DNA content of G_0/G_1 and G_2/M cells, then the data is plotted on a linear scale. If the range of fluorescence intensity is greater than one decade, then a four or five decade log scale is used to facilitate data display.

Color Compensation
Emission spectra of commonly used fluorochromes in flow cytometry can be broad and in multicolor flow cytometry they can often overlap. For example, although the emission peaks of fluorescein (FITC) and phycoerythrin (PE) are distinctly different, emission spectrum of FITC overlaps that of PE (Figure 2, spectral overlap indicated by the arrow). Therefore, it is important to remove the overlapping signal of FITC in the PE range particularly while using these two dyes in dual parametric analysis or co-localization studies. The process of subtracting the overlapping FTIC signal in the PE channel is called 'color compensation'.

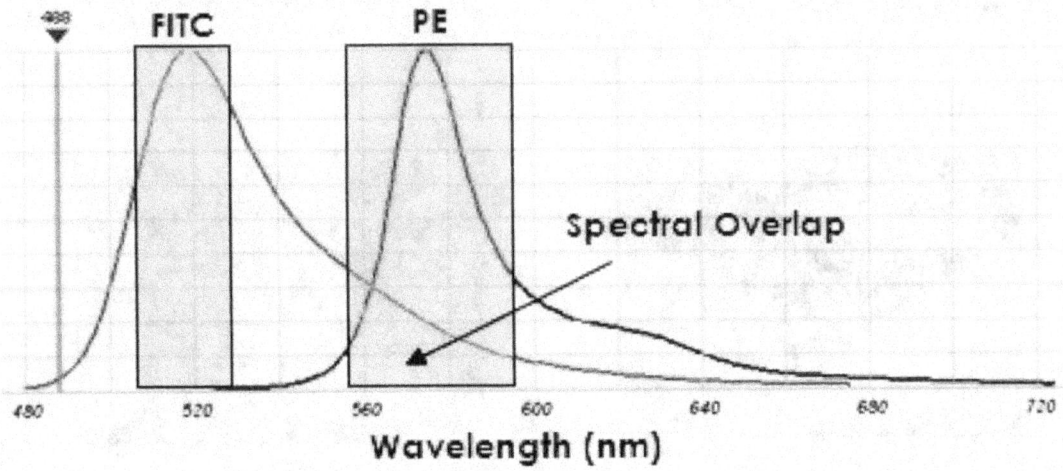

Figure 2: The emission spectra of fluorescein (FITC) and phycoerythrin (PE). The peak of emission of FITC and PE is collected by 530/30 and 585/42 band pass filters respectively (shown in light grey).

Color compensation is necessary particularly when acquiring data from multicolor flow cytometric analysis. Generally, the compensation is performed prior to the acquisition; however, digital flow cytometers will allow post-acquisition compensation. For example, to perform compensation while using FITC and PE stained samples, first we have to run a unstained sample and adjust the PMT voltage in such a way that the population is placed in the lower left quadrant of the dot plot and note the median of this population on

FITC and PE channels to set the compensation values of stained samples. Next the unstained and FITC stained sample is run and the FITC sample will be automatically placed in the upper right quadrant (Figure 3a) because of spectral overlap. Color compensation is performed by subtracting the percent of FITC fluorescence in the PE channel (i.e. PE - % FITC). The median of the unstained and FITC samples must be the same on the PE channel (Figure 3b). Similarly, run the unstained and PE sample (Figure 3C) and perform color compensation by subtracting the % of PE fluorescence in the FITC channel (i.e. FITC - % PE) to make sure that the median of the unstained and PE stained samples are the same on the FITC channel (Figure 3d).

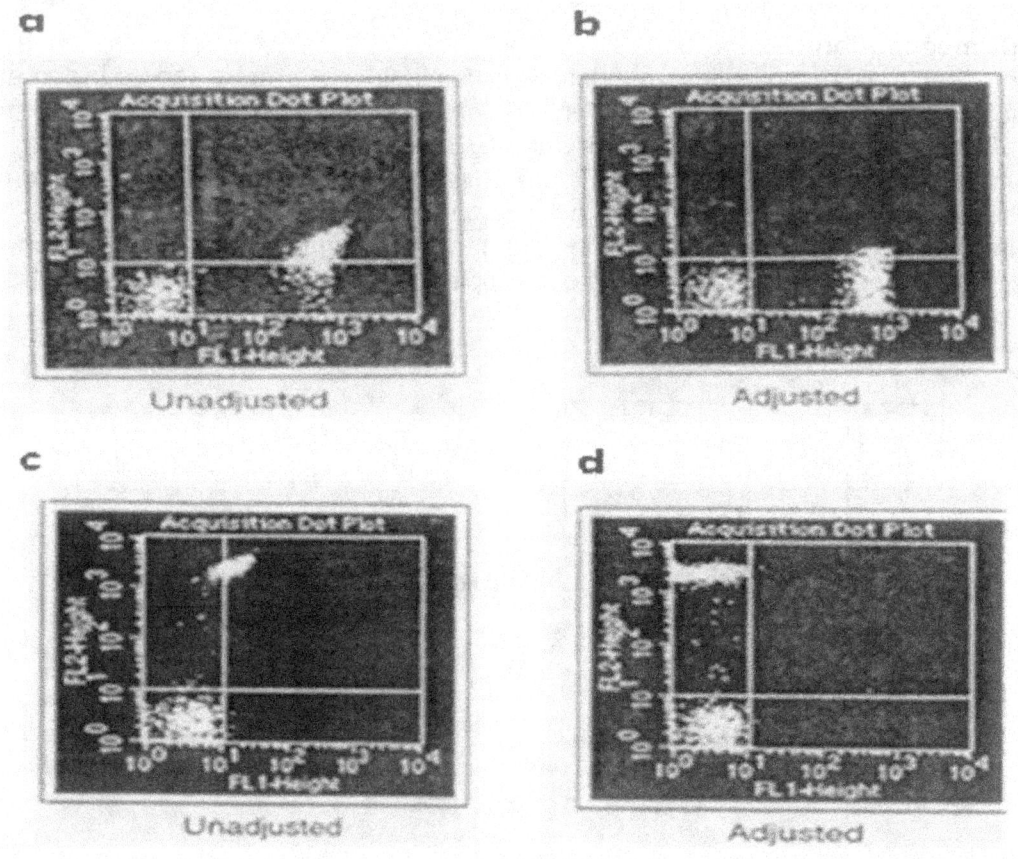

Figure 3: a) Unstained sample and FITC stained sample uncompensated, b) Unstained sample and FITC stained sample compensated, c) Unstained sample and PE stained sample uncompensated, d) Unstained sample and PE stained sample compensated.

Fluorescence Activated Cell Sorting
Fluorescence activated cell sorting is used to isolate and collect specific cells from a heterogeneous population. Although, there are several other methods available for cell separation (e.g., magnetic elution), fluorescence activated cell sorting is widely used because of the high yield and purity of the target population that can be obtained by this

procedure. In all stream-in-air sorters, the mechanism of sorting is based on the formation and deflection of droplets carrying a cell in the electrical field. The nozzle through which the cells are passed is about 5 to 7 times bigger than the cell size. The sorter that works at 70 PSI with a 70 μm nozzle will be vertically vibrated at a frequency of 20 to 70 kHz. The distance between the point of interrogation and break-off point is called the drop delay. The drop delay setting is critical for accurate sorting. When a target cell signal is acquired and identified by the electronic gate and it reaches the break-off point, the entire stream is charged and the droplet carrying target cell breaks-off and gets deflected towards the plate that has opposite charge and lands in a collection tube. The non-targeted cells and debris along with empty droplets will be aspirated into the waste tank. The fluidics stability and drop delay can depend on the temperature, humidity and air flow in the room and any changes in these ambient parameters will affect sort purity and yield.

References

Ormerod, M.G. 2008 Flow Cytometry - A Basic Introduction: this excellent book can be accessed and read for free from Denovo software site:
http://flowbook-wiki.denovosoftware.com/

Shapiro, H. 2003 Practical Flow Cytometry, 4th edition Wiley-Liss: this book can be accessed and downloaded from the Beckman Coulter flow cytometry page:
http://www.coulterflow.com/bciflow/practical.php

2. Sample Processing in a Clinical Cytometry Laboratory

Y. Badrinath
ybadrinath@rediffmail.com
A. Sehgal
kunalsehgal@gmail.com
A. Kumar
Ashok.san2010@gmail.com
P.G. Subramanian
pgs_mani@yahoo.com
S. Gujral
s_gujral@hotmail.com

Introduction
Flow cytometric (FCM) immunophenotyping in the clinical laboratory is a fast and objective method for the identification of hematolymphoid neoplasms. Additionally, it also provides information relevant for prognosis and therapy. Many advances made in instrumentation, software analysis tools, and in the availability of newer fluorochromes have facilitated use of multi-color immunophenotyping. We briefly discuss various pre-analytical variables in the analysis of hematolymphoid neoplasms.

Specimen collection
A variety of specimens are submitted for FCM analysis including peripheral blood (PB), bone marrow (BM), body fluids (pleural fluid, ascitic fluid, cerebrospinal fluid and pericardial fluid) and fresh tissue samples as lymph nodes (LN) aspirates/biopsies and other surgical materials.
Components involved in sample collection for FCM are adequately addressed in the literature. An important document is the clinical laboratory standards institute (CLSI) (formerly NCCLS) guideline for global application, developed through the clinical and laboratory standards institute consensus process. It is proposed that operators of clinical laboratories may use this document for information on sample collection.

Standard Precautions
All human specimens are to be treated as infectious and handled according to "standard precautions". Standard precautions are guidelines that combine the major features of "universal precautions and body substance isolation" practices. Standard precautions cover the transmission of any pathogen and thus are more comprehensive than universal precautions which are intended to apply only to transmission of blood-borne pathogens.

Specimen labeling and test ordering
Good clinical practice dictates that specimen containers be labeled immediately at the point of collection by the collector (e.g., phlebotomist at the bedside). The label should include at least two unique patient identifiers (e.g., full patient name, medical record

number, or patient birth date). Ideally, patients (if capable) should read back each label's patient identifiers to the phlebotomist or sign each label to attest to the accuracy of the identifiers. Samples should be submitted with the formal order. The latter (paper or electronic) should include time and date of collection and, preferably, the name of the phlebotomist or physician obtaining the samples.

Trained staff (Cytometrists/technologists, hematopathologists)
It is crucial to have written guidelines for technical staff (e.g., histo-technologists, cyto-technologists, and medical technologists) and professional medical staff (e.g., pathologists, surgeons, radiologists, and residents). In particular, there should be documentation of training of such individuals and annual competency testing. Such instructions also should be included in the laboratory user guide.

Equipment
- Flow cytometer
- Table top centrifuge with swing out rotor
- Vortex Mixer
- Plastic racks for glass tubes
- Polystyrene round bottom tubes for assays and acquisition
- Micropipettes 0.5-10 µl, 10-100 µl, 200-1000 µl (and tips)

Reagents
- Phosphate Buffered Saline (pH 7.2) with 2% Bovine Serum Albumin (PBS-A)
- RBC Lysing Solution (NH_4Cl based or commercial lysing solutions)
- Fluorescent Conjugated Antibodies
- Quality Control Beads (eg: Calibrite Beads/CST beads (cytometer set-up and tracking beads)/ Flowcheck beads)
- Sheath Fluid
- 5% Hypochlorite Solution or Commercial cleaning solutions
- Labeled antibodies

Specimen anticoagulation
1. **Bone marrow or Peripheral blood**
   a) **Immunophenotyping**
      Anticoagulants are required for liquid hematology samples. These include EDTA, sodium or lithium heparin, and acid citrate dextrose (ACD). Typically, EDTA is used for PB samples and heparin for BM. However, any of the above three anticoagulants are acceptable for PB or BM samples submitted for the immunophenotypic evaluation of hematolymphoid neoplasms. All three anticoagulants preserve the clinically relevant antigen expression for traditional leukemia/lymphoma immunophenotyping for 24 to 48 hours, with ACD probably having the longest shelf life (approximately 72 hours). This should allow time for some degree of morphologic triage (eg. Wright's-stained aspirate smears or core biopsy imprints), if desired. The preference of

heparin anticoagulation for BM may be enhanced in the era of FCM analysis of myeloid antigen expression for myelodysplastic syndromes (MDS). Granulocytes begin to show significant apoptosis after 6 hours in EDTA or ACD anticoagulation, with heparin showing the least amount of apoptosis. This has significant impact on light scatter for MDS assessment and oxidative burst for chronic granulomatous disease evaluation.

b) **Quantitative Blood Lymphocyte Subset Analysis**

EDTA or heparin collection tubes are indicated for quantitative blood lymphocyte subset analysis. An ACD tube should not be used because of the 18% volume dilution by the liquid anticoagulant. This, combined with inconsistent filling of the tube by a phlebotomist, results in an unacceptable level of variation in the in vitro concentration of leukocytes and it is formally discouraged in some recommendations. Although forward light scatter is maintained in EDTA for only 24 hours, the widespread adoption of CD45 versus side light scatter gating for lymphocyte quantitation has allowed the shelf life for EDTA anti coagulated samples to be extended to 72 hours.

c) **CD34 hematopoietic stem cell quantitation**

For CD34 stem cell quantitation, an EDTA anti coagulated specimen is recommended for single- or dual-platform techniques. This choice and the avoidance of ACD are for the same reasons as for lymphocyte quantitation. However the actual pheresis product for CD34 stem cell collection usually is already anti coagulated with ACD. Thus, an aliquot simply can be collected in a sealed clean container and sent to a laboratory.

2. **Fresh Tissue Specimens**

In general, anticoagulant is not required for fresh tissue samples. These may be transferred to a transport media (see below).

3. **Body Fluids**

For body fluids (pleural fluid, ascitic fluid, and cerebrospinal fluid, pericardial) anticoagulation is not required unless unless the fluids are grossly hemorrhagic in nature. One can simply collect samples in a clean sterile container.

Specimen Age and Integrity

One of the salient questions in clinical flow immunophenotyping is how long samples submitted for hematolymphoid neoplasms analysis may be held?

Because specimen age is a critical variable in FCM analysis, specimens must be labeled with the date and time of collection and processed as soon as possible. This is particularly important in cases of tumors with high proliferation rate (e.g. Burkitt's lymphoma) or in patients where chemotherapy and/or radiation treatments were given. A 48 hour cut-off for specimen age is appropriate; however, irreplaceable specimens should not be rejected if they exceed the 48 hour specimen age requirements. Every attempt should be made to derive useful information from such samples. A statement regarding exceeded specimen age should be noted in the final report. Some antigens are more labile than others (e.g., CD138) and may be preferentially affected with

increasing specimen age. The potential false negativity of labile antigens should be noted when interpreting the antibody staining patterns in older specimens.
Although the peer-reviewed literature is limited, there is a fair degree of unpublished experience that allows for reasonable guidelines.

Specimen Requirements
Various types of specimen used for immunophenotyping include:
1. Peripheral Blood in 2% EDTA vaccutainers (Lavender Cap). Sample best processed within 24 hours and results can be erroneous after 3 days
2. Bone Marrow in 2% EDTA vaccutainers (Lavender Cap). Sample best processed within 24 hours and results can be erroneous after 3 days
3. Body Fluids in Plain vaccutainers (Red Cap): For grossly hemorrhagic body fluid or aspirates 2% EDTA vaccutainers are preferred to prevent sample from clotting. Cerebrospinal fluid is collected as such. If hemorrhagic, it may be collected in EDTA vaccutainer (Lavender Cap).
4. Fine Needle Aspirates in normal saline in plain vaccutainers (Red Cap) or any other containers
5. Fresh Lymph Node Biopsies in Normal Saline and are processed on urgent basis.

Storage conditions:
Samples once collected should be sent to the flow cytometry laboratory as a soon as possible. Samples should be stored at ambient room temperature (22-24 degree centigrade), upto 24 hours. Beyond 24 hours refrigeration is recommended to avoid sample degeneration. Body fluids, fine needle aspirates and suspected Burkitt's lymphoma cases should be processed immediately as these samples degenerate very fast. In cases where a delay is anticipated, the body fluids, fine needle aspirate as well as lymph node biopsy should be collected in a transport media (RPMI with fetal calf serum).

1. **Bone marrow or Peripheral blood for immunophenotyping**
   Liquid hematology samples for leukemia/lymphoma evaluation usually can be stored at room temperature for 24 to 36 hours without loss of significant antigenicity. This typically can be expanded to 48 to 72 hours if such samples are diluted 1:1 with cell media (e.g. RPMI-1640 or McCoy's minimal media) and stored at 4°C.
2. **Quantitative Blood Lymphocyte Subset Analysis**
   Blood samples for quantitative lymphocyte subset analysis can be held for up to 72 hours in EDTA anti-coagulated containers without clinically significant changes in the concentrations or percentages of lymphocytes but only if CD45 versus side light scatter gating is used. If forward light scatter is a component of the lymphocyte gating, then EDTA anti-coagulated samples typically have a 24-hour limit before significant loss of forward light scatter is seen on lymphocytes. Because of this and the overall superiority of CD45 versus side light scatter gating, the use of forward light scatter in lymphocyte gating is not recommended.

Quantitative lymphocyte analysis was performed initially with density gradient purified samples. The CD4 T-cell subset was vulnerable particularly to low-temperature–induced loss during density gradient purification (so-called 'refrigerator AIDS' or "RAIDS"). This issue essentially has disappeared with the switch to whole blood lysis methods, which allow blood samples to be held at room or refrigerated temperatures without loss of accurate gating. Transport and storage conditions should assure, however, that samples are not exposed to freezing or excessively warm conditions (e.g., 37°C).

3. **Solid Tissues**
   Ideally, solid (excisional) biopsy tissue samples should be processed immediately (1 hour after biopsy) into a single-cell suspension. This can be done by the classic "mince and slice" technique, with or without the use of a steel mesh screen. The crude suspension should be placed in a 1:1 dilution of cell media (e.g. sterile chilled RPMI-1640 or McCoy's minimal media) and stored at 4°C until a decision is made whether or not to immunostain the sample. In the authors' experience, there is a high degree of confidence in holding such samples for 24 to 48 hours, particularly in the differential diagnosis of reactive versus low-grade B-cell lymphoid lymphomas.
   Again, although an extensive body of literature is not available, it is possible (but not ideal) to hold suitably prepared solid tissue for several hours (up to 24 hours) and still have a reasonable degree of confidence that clinically relevant immunophenotyping data can be obtained. To realize this, samples should be small (1 cm in greatest dimension) and thin (0.5 cm). Such slices should be wrapped in saline-soaked gauze, and then immersed in sterile saline in a sealed container stored at 4°C. At the earliest opportunity, samples should be processed into a crude cells suspension (discussed previously) or a decision made whether or not to perform immunophenotyping.

4. **Body Fluids**
   If the specimen is to be held for any period of time, viability may decrease if tissue culture media is not used as a transport media. However, auto fluorescence may increase with the use of tissue culture media and consideration should be given to the impact of low viability vs. auto fluorescence when choosing a transport media. Cold (4 °C) storage may be useful for prolonged storage of some specimens such as pleural fluids and cell suspensions prepared from tissues. Extreme temperatures must be avoided.
   When specimen analysis is delayed by shipment to a remote laboratory, a freshly prepared and stained smear or tissue imprint and/or tissue section should accompany specimens for reference. An alternative is the addition of a stabilization fluid to the sample; this will allow specimen storage for up to 10 days so that retrospective analysis can be undertaken.

Processing of Samples
The order followed for processing of hematolymphoid samples for immunophenotyping starts with lysing (red cell) of the sample, followed by wash and then staining (lyse-wash-

stain method). The first step is to obtain a single cell suspension followed by antibody staining, data acquisition and data analysis.

Lysing is usually used for Peripheral Blood or Bone Marrow samples or hemorrhagic body fluids/aspirates (to get rid of red cells in case of immunophenotyping).

1. **Cell suspension preparation**
   a) **Peripheral Blood (PB)/ Bone Marrow(BM)**
      A ratio of 1:10 of sample to lysing solution is recommended. (Mix 0.5ml of the sample with 4.5 ml of lysing solution in a test tube. Incubate for 6-8 minutes at room temperature. Centrifuge for 2 minutes at 400xg at room temperature to remove red cell debris. Discard the supernatant. Gently tap the pellet to resuspend the cells and add 2 ml of PBS-A. Wash by centrifuging for 2 minutes at 400xg at room temperature to remove the left over lysing solution. Discard the supernatant and re-suspend in PBS-A to get the final cell suspension. Target cell count – 0.5 to 2 x10^6 cells per 50 µl per assay tube

   b) **Body Fluids / Fine Needle Aspirates**
      Centrifuge (400xg at room temperature for 2 minutes in swing rotor centrifuge) to obtain a cell pellet. Gently tap the pellet to resuspend the cells and add 2 ml of PBS-A. Wash by centrifuging for 2 minutes at 400xg at room temperature. Discard the supernatant and re-suspend in PBS-A to get the final cell suspension. Target cell count – 0.5 to 2 x10^6 cells per 50 µl per assay tube

      As body fluids and aspirates may have low cell yields one may have to reduce the target cell count to as low as 25000 to 50000 cells per tube and select priority tubes from a given panel. Optimal acceptable percent of viable cells is about 80%.

      *For Hemorrhagic fluid samples, a lysing step may be required (as described above for PB/BM).

   c) **Lymph Node Biopsy/Tissue Biopsy in Normal Saline**
      Suspend the biopsy in normal saline and transfer to a small Petri Dish. Cut the biopsy into tiny bits and gently tease with the help of scissors and forceps. The cells can be transferred from the petri dish to a test tube with micropipettes. Centrifuge the sample to obtain a cell pellet (400xg at room temperature). Gently tap the pellet to resuspend the cells and add 2 ml of PBS-A. Wash by centrifuging for 2 minutes at 400xg at room temperature. Discard the supernatant and re-suspend in PBS-A to get the final cell suspension. The target cell count is around 0.5 to 2 x10^6 cells per 50 µl per assay tube.

2. **Staining**
   Antibody staining is performed according to the manufacturer's instructions. After the final cell suspension is ready and target cell concentration is obtained, antibodies can be added as per the lymphoma panels of the laboratory (based on history, clinical details, morphology and other findings or else as per the policy of

the laboratory). As mentioned above, for low cell yields samples (e.g., cerebrospinal fluid) one may have to select priority tubes from a given panel. A given panel may have four to six tubes per panel with each tube containing three to six antibodies (colors) based on cytometer configuration. One can add individual antibodies or pre-prepared antibody cocktails based on the sample work load and the laboratory policy.

Add 50 µl of cell suspension to individual antibodies/antibody cocktail (the volume of antibody added is based on the titration volumes determined for each antibody). Mix gently by tapping and incubate for 15-20 minutes at room temperature in the dark. Wash with 3-5 ml of PBS-A by centrifuging for 2 minutes at 400xg at room temperature to remove excess unbound antibodies.

Discard the supernatant and re-suspend in 0.5 ml of 0.5% paraformaldehyde. For cytoplasmic and nuclear staining, cells are fixed and permeabilized before antibody staining and such reagents may be prepared in-house or procured commercially. After staining, the cells are run on the flow cytometer, data stored and analysed and a specimen report created. Turn around time for reporting a case of leukemia/lymphoma immunophenotyping may be from 1-2 days. A stringent internal as well as external quality control program for lab operation should be maintained.

References

Clinical and Laboratory Standards Institute. Clinical Flow Cytometric Analysis of Neoplastic Hematolymphoid Cells; Approved Guideline, 2nd ed. CLSI document H43-A2. Wayne, PA: Clinical and Laboratory Standards Institute; 2007.

Clinical and Laboratory Standards Institute. Enumeration of Immunologically Defined. Cell Populations by Flow Cytometry; Approved Guideline, 2nd ed. CLSI document H42-A2. Wayne, PA: Clinical and Laboratory Standards Institute; 2007.

Clinical flow cytometric analysis of neoplastic hematolymphoid cells, approved guideline-sections 6 –9, Clinical lab Med 27 (2007) 607 – 707.

Infection Control and Hospital Epidemiology. 1996; 17(1):53-80), MMWR 1987; 36 [suppl 2S] 2S-18S), and (MMWR 1988; 37:377-382, 387-388).

3. Flow Cytometric Analysis of Lymphocyte Subsets (Immunophenotyping)

Raquel Cabana
rcabana@blueoceanbio.com

Introduction

The objective of immunophenotyping is to enumerate the antigenically defined three major lymphocyte populations of T lymphocytes (CD3+), B lymphocytes (CD19+) and Natural Killer lymphocytes (CD56+/CD3-). T lymphocytes can be subdivided into CD3+CD4+ T helper lymphocytes and CD3+CD8+ T cytotoxic lymphocytes. It is important to determine the relative and absolute numbers of these subsets to determine if a sample is abnormal (Clinical and Laboratory Standards Institute). Changes in lymphocyte subsets, such as T or B cell populations, can be indicative of immunological changes related to various diseases. The importance of monitoring the immune system became especially important in monitoring of human immunodeficiency virus (HIV) infection where alterations in peripheral blood CD4+ T cell content can be used for diagnosis and prognosis of HIV infection and the management of patients receiving antiretroviral therapy (Bergeron et al., 2002).

Preparation of samples for phenotype analysis is dependent on the type of flow instrument available, the antibodies and the fluorescent conjugates selected and the use of lysing agents. Sample preparation can be performed manually by pipetting or automatically in a sample prep station when a large number of samples have to be processed and analyzed. Recently, Blue Ocean Biomedical, Inc. has announced availability of an integrated system which performs automatic sample preparation, data acquisition and analysis on a single platform.

Methods for Cells Enumeration

Historically, lymphocytes analysis was performed on a white blood cell buffy coat sample collected by density sedimentation. However, the need for accurate enumeration has replaced the density gradient method by the whole blood lysis procedure. The absolute cell concentration is defined as the number of cells of interest in a known volume of the sample using either the Single or the Dual Platform methods (Clinical and Laboratory Standards Institute, Bergeron et al., 2003). The Single Platform bead method is based upon the addition of a known number of fluorescent microspheres to a known volume of sample, either using a tube containing a known number of fluorescent microspheres to which sample is added or by the addition of a specific volume of a calibrated bead suspension to the sample tube. Fluorescent beads of a known concentration can be obtained from a number of different manufacturers. The use of this counting option has significantly reduced the variability of data between labs. In the Single Platform volumetric method, an accurate and reproducible volume of stained sample is passed through the flow cell, and the number of cells analyzed can be directly related to the known volume of the sample analyzed. The sample volume can be measured by using either precision syringes or sensing electrodes in the flow analyzer. The pipetting steps in volumetric systems must be performed with great precision. The final dilution of the sample must also be calculated and taken into account. In the Dual Platform method, two instruments, a hematology analyzer and a flow cytometer are used to count the cells of interest as a percentage of a reference population (e.g. CD3+/CD4+ cells as a percentage of the total lymphocyte population). Then an absolute WBC count provided by a hematology analyzer is used to calculate the percentage of lymphocytes. A disadvantage of the dual platform method is that errors could be compounded by the use of two different instruments.

General Specimen Information

- **Instrument QC -** Instrument setup and performance qualification are instrument specific. CLSI recommends daily performance monitoring of fluorescence intensity, color compensation and verification of system performance using QC controls. For this purpose, one can use commercially available preserved cell preparations which express most of the phenotypic markers and can be processed along with the blood sample to be analyzed.

- **Specimen Collection -** Samples must be collected in Vacutainer™ tubes containing either EDTA (lavender top) or sodium heparin (green top). If WBC counts and differential are obtained from the same specimen, the anticoagulant of choice is EDTA. Samples In sodium heparin are stable for up to 72 hrs while in EDTA they are stable up to 48 hrs. EDTA samples will become depleted of granulocytes after 24 hrs; however CD4 counts are still possible up to 120 hrs when using optimized lymph gating strategy.

- **Specimen Rejection -** Blood samples with any of the following characteristics may not yield reliable data and thus should not be processed:
  1. Clotted Samples
  2. Samples collected in an improper tube
  3. Partial blood draw
  4. Insufficient number of viable cells (a minimum of 75% viability is recommended).
  5. Samples exposed to temperature extremes
  6. Samples Improperly labeled
  7. Loss of specimen integrity (e.g., Lipemia)

- **Selection of Lysing Agent -** After immunostaining of the cells, RBCs lysis is used to remove erythrocytes from the sample. Several lysing methods are available for whole blood lysis and the mechanism of action for red blood cell lysis depends on the nature of the reagent. Ammonium chloride is the most popular lysing reagent which increases the osmotic pressure resulting in a rupture of the red cell membrane. Other lysing mechanisms include incubation of cells with formic acid or other acidic solutions followed by addition of a basic agent to neutralize pH or the use of diethylene glycol or detergents such as saponin (Bossuyt et al., 1997).

- **Immunostaining Panel** - The ability to measure multiple cell surface markers is limited by the number of fluorochromes that can be simultaneously detected in a particular instrument. The fluorochromes selected are determined by excitation source and the wavelengths for excitation and the emission filters available. When building a marker panel, it is important to choose the brightest fluorochrome for the least expressed protein (e.g. PE for CD56) and the dimmest fluorochrome for the most expressed protein (e.g. FITC for CD45). It is also essential to choose fluorochromes with the least possible spectral overlap to minimize the need for color compensation. Tandem dyes should be used with caution as they may suffer from uncoupling and photo-bleaching. Color compensation should be performed as part of the instrument set up before sample analysis to avoid spectral overlap between the different fluorochromes. The optimal method for setting up compensation matrix is to use cells stained with a single color immunofluorescent marker (Cabana et al.)

- **Principles of the procedure** - The fluorochrome-labeled antibodies are mixed with the sample where they bind to the cell surface antigens. The stained samples are then incubated with an erythrocyte lysing solution to remove the RBCs before analysis. If samples cannot be analyzed immediately, a fixative can be added to preserve the stained and lysed samples.

- **Reference Intervals** - Reference intervals for immunophenotyping test results must be determined for each laboratory. Separate reference intervals must be established for adults and children (CLSI guideline C28). The following table summarizes the mean ± SD, median and 95% reference range of percentages and absolute values of lymphocyte subsets in 220 healthy adults (Yaman et al., 2005).

Lymphocyte Subsets		Mean ±SD	Median	95% reference range
CD3	%	72.70±8.44	73.20	52.30-84.64
	cells/µL	1680±528	1637	725-2960
CD4	%	47.37±9.10	49.35	30.00-60.34
	cells/µL	1096±391	1055	437-2072
CD8	%	28.99±5.99	29.65	17.76-39.94
	cells/µL	669±239	630	307-1184
CD19	%	10.96±4.44	10.85	3.90-20.79
	cells/µL	254±122	240	74-586
CD56	%	7.03±3.26	7.15	0.10-13.20
	cells/µL	161±92	155	3.08-367
CD4/CD8		1.68±0.43	1.65	1.06-2.76

Lymphocyte Subset Analysis

The following figures show a "Four-Color, Dual Anchor" gating strategy to identify the major lymph subsets (T, B and NK) while providing several internal quality controls.

Tube 1: CD45- CD4-CD8-CD3. (Count helper/cytotoxic T cells)
Tube 2: CF45-CD56- CD19-CD3. (Count total TBNK)

In the examples, the Blue Ocean Biomedical LSA panels include the following antibodies:
> LSA1: CD45 FITC, CD4 PE, CD8 PE DL594, CD3 PE DL649.
> LSA2: CD45 FITC, CD56 CD16 PE, CD19 PE DL594M CD3 PE DL649.

The advantages of this panel are:
1. lymphocytes are easily distinguished based on CD45 fluorescence and 90' (side) scatter
2. Replicate CD3 determinations ensure reproducibility between tubes.

Four Color; Two tube Lymphocyte Subset Enumeration Assay

Figure 1. Whole Blood Stained with LSA1 CD45/CD4/CD8/CD3 and Analyzed on a Blue Ocean Biomedical CR150 flow cytometer.

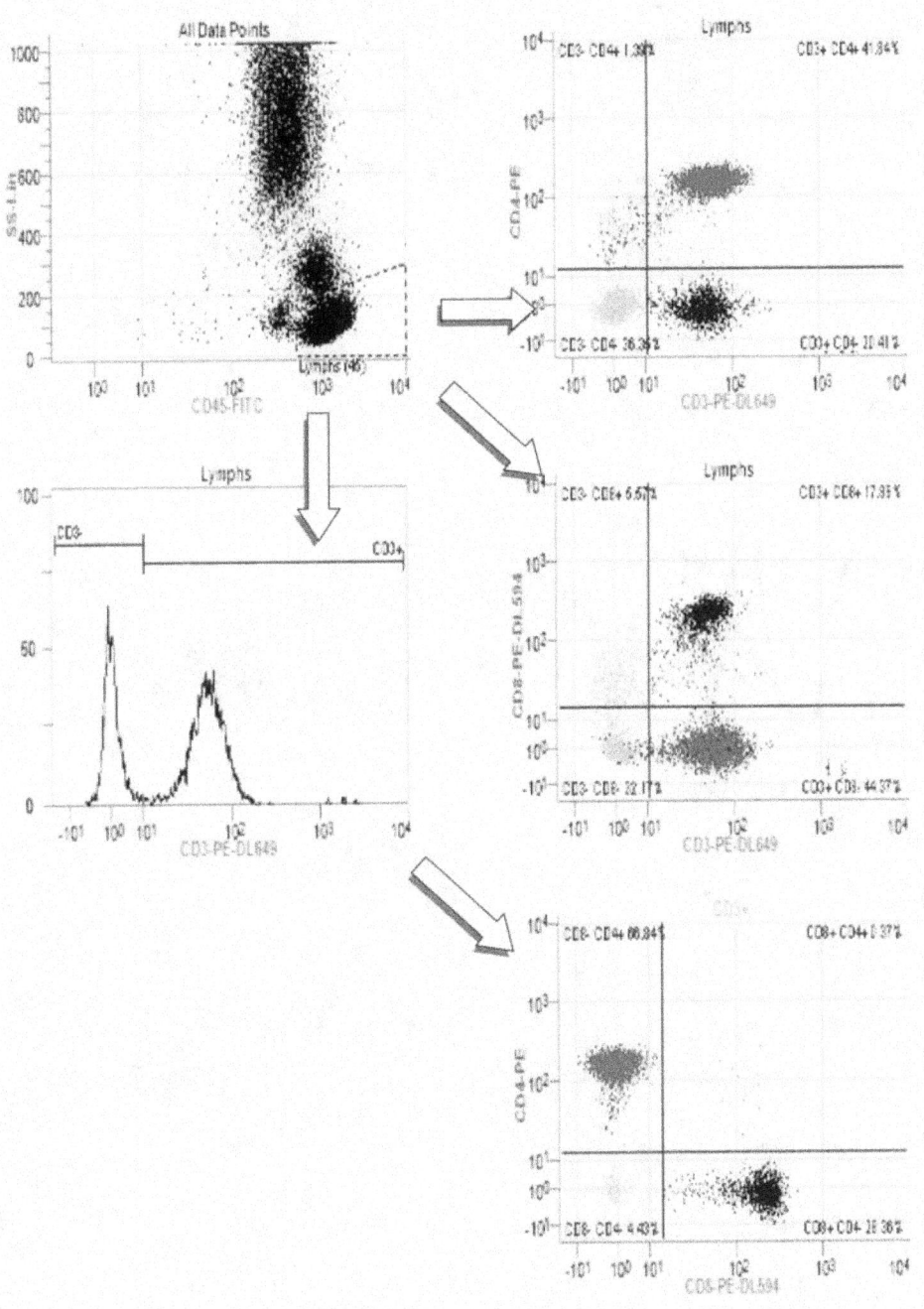

Figure 2. Whole Blood Stained with LSA2 CD45/CD56CD16/CD19/CD3 and Analyzed on a Blue Ocean Biomedical CR150 flow cytometer.

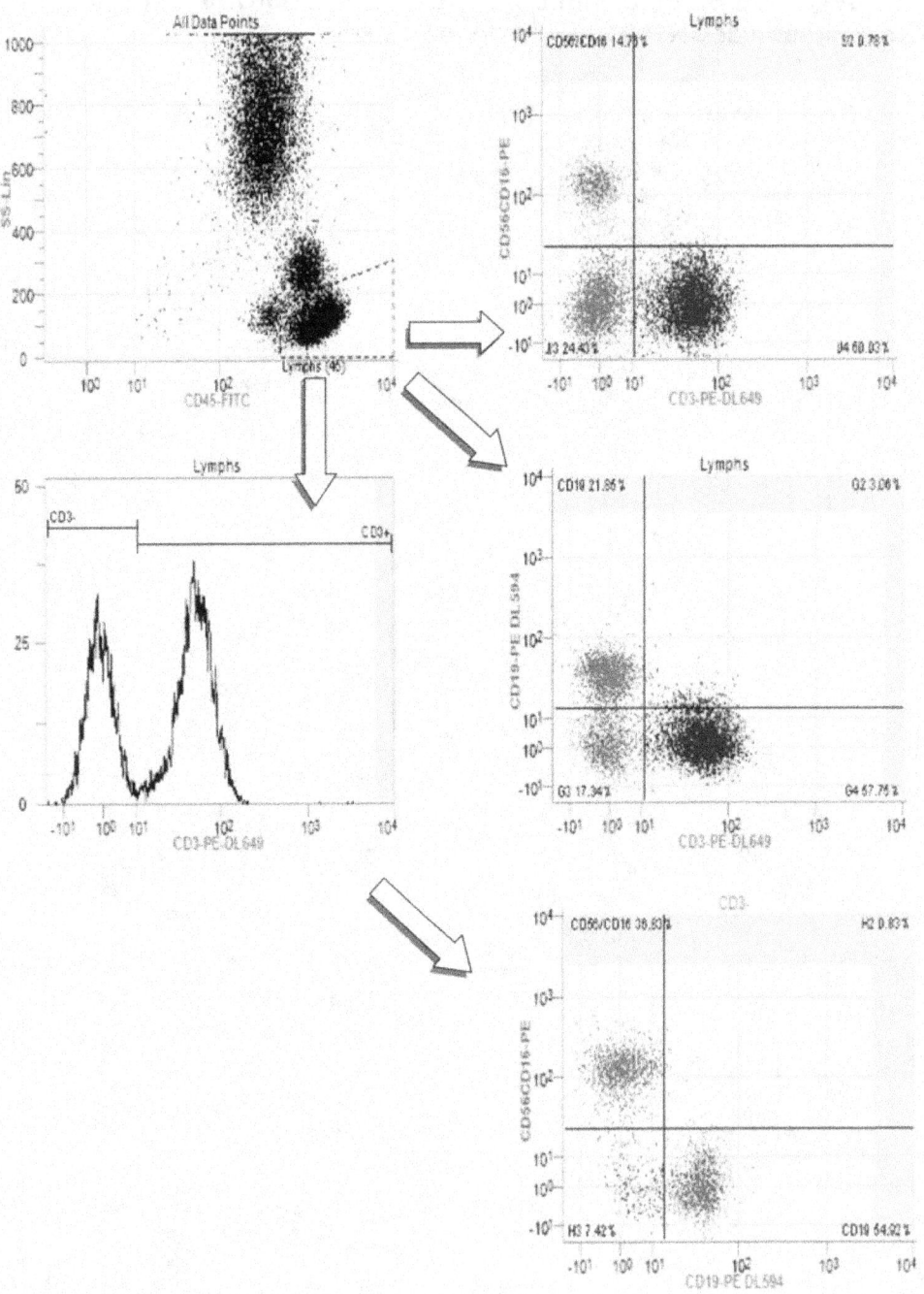

Acronyms, Definitions, and Abbreviations:

Ethlylene diaminetetraacetic acid	EDTA
Dual Platform	DP
Single Platform	SP
RBCs	Red Blood Cells
HIV	Human Immunodeficiency Virus
MABs	Monoclonal Antibodies
LSA	Lymphocyte Subset Analysis
WBCs	White Blood Cells
Hrs	Hours
CD	Cluster of differentiation.
FITC	Fluorescein isothiocyanate
PE	Phycoerythrin
QC	Quality Control
PE DL	Phycoerythrin Dy Light
CLSI	Clinical and Laboratory Standards Institute

References

Bergeron M, Lustyik G, Ding T, Nicholson J, Janossy G, Shapiro H, et al. When do non volumetric flow cytomters become volumetric? Time can tell .how absolute your instrument is about absolute counts. Cytometry 52B:37-39 (2003).

Bergeron M, Nicholson J, Phaneuf S,Ding T, Soucy N, Badley A, Hawley Foss C, Mandy F. Selection of Lymphocyte Gating Protocol Has an Impact on the Level of Reliability of T-Cell Subsets in Aging Specimens. Cytometry (Clinical Cytometry) 50:53–61 (2002).

Bossuyt X, Marti G, Fleisher T. Comparative Analysis of Whole Blood Lysis Methods for Flow Cytometry. Cytometry (Communications in Clinical Cytometry) 30:124–133 (1997).

Cabana R, Cheetham M, Enten J, Song Y, Thomas M and Brendan SY. 3-Color Compensation. Beckman Coulter Application Information. A2045A.

Clinical and Laboratory Standards Institute - Enumeration of Immunologically Defined Cell Populations by Flow Cytometry; Approved Guideline – Second Edition, document H42-A2.

Yaman A, Cetiner S, Kibar F, Tasova Y, Seydaglu G, Dunbar I. Reference Ranges of Lymphocyte Subsets of Healthy Adults in Turkey. Med Princ Pract 14:189–193 (2005).

4. Propidium Iodide Staining for DNA Content And Cell Cycle Analysis

Awtar Krishan
akrishan@med.miami.edu

Introduction

Crissman and Steinkamp (1973) published a paper on use of propidium iodide for flow cytometric analysis of DNA in cells fixed with ethanol and digested with RNase. Subsequently, Krishan (1975) reported that propidium iodide solutions made in hypotonic sodium citrate could lyse the cells and directly stain the cells for DNA content and cell cycle analysis. The method described below is rapid, uses small amount of material and has been extensively quoted and used for flow cytometric analysis of DNA aneuploidy and cell cycle distribution.

Reagents:

- Propidium iodide – Sigma P-4170
- Trisodium citrate dihydride – Sigma S-4641
- Nonidet P-40 (NP-40) – Sigma N-6507
- NP-40 equivalent – Sigma I-3021 Igepal CA-630
- Tween 20 – Sigma P-1379
- Pepsin – Sigma P-7012 from porcine stomach
- RNase A – Sigma R-4875
- Phosphate buffer saline (PBS)
- 37 µm Nylon mesh – Small Parts Inc. CMN-0040-D

Solution for live cells

- Hypotonic propidium iodide, 50 µg/ml in 0.1% trisodium citrate dihydride containing 0.3 µl/ml of Nonidet P-40 or a non-ionic detergent equivalent (Igepal CA-630). Solution should be stored in a dark amber bottle in a refrigerator (4-8 °C) and will keep for years.

Solutions for formalin fixed/paraffin embedded tissues

- Isotonic propidium iodide, 50 µg/ml in phosphate buffered saline (pH 7.0-7.5) containing 2 mg/ml of RNase A. Solution should be stored in a dark amber bottle in a refrigerator (4-8 °C) and will keep for years.
- Trisodium citrate dihydride, 3 gm/L with 0.05% Tween 20 (Adjust to pH 6.0 with 1 N HCL.) This solution is for antigen retrieval.
- Pepsin, 0.05% pepsin in a 0.85% NaCl solution (Adjust to pH 1.00 with 1 N HCl).

Method
Buffy coats from Ficol-Hypaque treated peripheral blood or bone marrow
1. Add approximately 10^6 cells/ml directly to the hypotonic propidium iodide stain and vortex vigorously.

Suspension cell cultures
1. Centrifuge the cells to collect a cell pellet.
2. Remove all medium by inverting and holding the test tube over a paper towel to drain.
3. Add hypotonic propidium iodide stain and vortex vigorously.

Monolayer cultures
1. Remove all medium and rinse the flasks or Petri dishes with phosphate buffered saline (PBS).
2. Add the hypotonic propidium iodide citrate stain to the monolayer and scrape with a rubber policeman, shake vigorously and use a fine tipped pipette to dislodge any cytoplasmic remnants from the lysed cells.
3. Centrifuge and suspend the nuclear pellet in fresh hypotonic propidium iodide citrate stain for analysis. Samples can be stored for up to 24 hrs in a refrigerator before flow analysis. The isolated nuclei will need more centrifugal force than cells and we normally use approximately 100 RCF.

Comments
- Do not use trypsin to remove the monolayer as it may interfere with dye binding.
- If floating cells in the monolayer culture are of interest, collect them by centrifugation and stain the pellet using the hypotonic PI solution.

Fresh or cryopreserved solid tumors and tissues
1. Remove necrotic and fat tissue.
2. Add hypotonic propidium iodide citrate stain to cover the tissue.
3. Mince with crossed scalpels or fine scissors. Use fine tipped pipette to mix vigorously.
4. Filter through 37 μm mesh nylon cloth to remove clumps.
5. Collect the cells by centrifugation at 100 RCF for 5 mins and resuspend in fresh stain for analysis.

Paraffin embedded formalin fixed tissues:
1. Cut 50 micron sections, transfer to a 5 ml glass tube. (Need to use glass tubes and glass Pasteur pipettes because the solvents will dissolve plastic and plastic tubes will not withstand the high temperature water bath during antigen retrieval.)
2. Remove paraffin by washing the sections 3X in 3 ml xylene or other solvents (e.g., Citrisolv, AmeriClear) for 10 min, each.
3. Rehydrate the sections by washing the sections in 3 ml of a descending ethanol series (100%, 75%, 50% 30%) for 5 min each. NOTE: It is very important when doing the first wash in 100% ethanol to remove ALL the xylene/solvent from the

inside of the tube. If any xylene/solvent remains in the tube and contacts the diluted ethanol washes, it will turn them milky and one will have to restart with the 100% ethanol wash.

4. Decant and wash the sections once in 3 ml distilled or deionized water.
5. Decant and wash the sections once in 3 ml phosphate buffered saline (PBS).
6. Decant and add 3 ml of the sodium citrate antigen retrieval solution.
7. Incubate for 30-60 min at 80°C in a water bath with intermittent shaking or pipetting every 5 min. NOTE: You do not want a temperature gradient to form as this will give uneven antigen retrieval and consequently uneven DNA staining.
8. Cool the samples at room temperature for 10 min and carefully decant the citrate.
9. Add 3 ml of the pepsin solution and incubate for 30 min at 37°C in a water bath. Vortex or briskly pipette the sample to break up the tissue and release the nuclei.
10. Centrifuge (100 RCF) and carefully remove the supernatant.
11. Resuspend in 3 ml PBS, centrifuge and carefully remove the supernatant.
12. Stain with nuclear pellet in isotonic propidium iodide/RNase solution at 37°C for 30 min.
13. Filter through 37 µm nylon mesh before analysis.

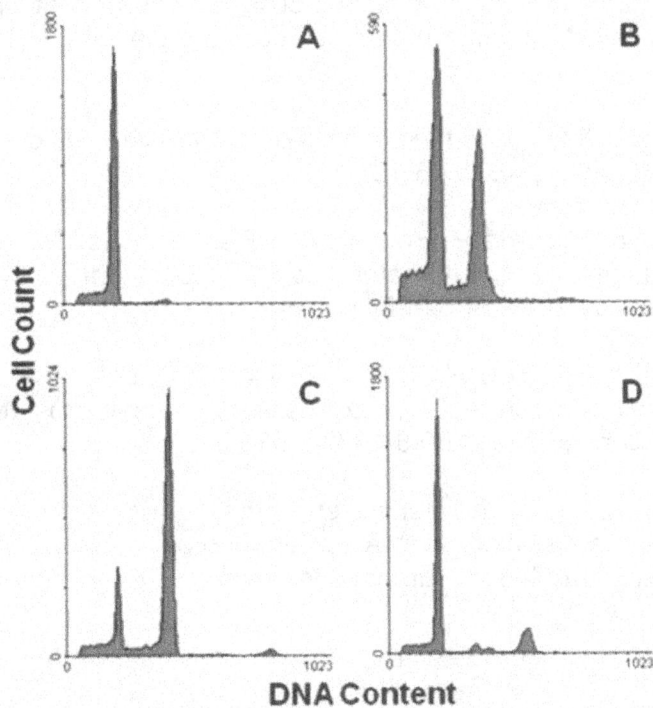

Figure 1 shows four DNA Distribution histograms of nuclei prepared after pepsin digestion from sections of formalin fixed-paraffin embedded breast tumors. Histogram 1A shows a predominant diploid population while histograms 1B and 1C show the presence of aneuploid (near triploid) besides the diploid peak. In histogram 1D, two smaller aneuploid populations with triploid and tetraploid DNA content are present besides the diploid peak.

Comments

- Propidium iodide is a DNA binding dye and could be mutagenic and thus use with caution.
- The stain can be stored at room temperature or in a refrigerator in a dark bottle and will keep for a long time.
- Sections in citrate solution for antigen retrieval should be gently mixed to avoid different temperature zones which may lead to uneven antigen retrieval.
- Propidium iodide concentration can be from 10-50 µg/ml and cell concentration should be approximately 10^6/ml.
- For dual staining with FITC labeled antibodies, the PI concentration should be halved.

References

*Adiga SK, Andritsch I, Rao RV, Krishan A. Androgen receptor expression and DNA content of Paraffin-embedded archival human prostate tumors. Cytometry 2002 50:25-30.

Crissman HA, Steinkamp JA. Rapid, simultaneous measurement of DNA, protein,and cell volume in single cells from large mammalian cell populations. J Cell Biol. 1973 59:766-71.

Krishan A. Rapid flow cytofluorometric analysis of mammalian cell cycle by propidium iodide staining. J.Cell Biology 1975 66:188-193.

*Krishan A, Ganjei-Azar P, Jorda M, Hamelik RM, Reis IM, Nadji M. Detection of tumor cells in body cavity fluids by flow cytometric and immunocytochemical analysis. Diagnostic Cytopathology 2006 34:528-541.

*Krishan A, Oppenheimer A, You W, Dubbin R, Sharma D, Lokeshwar B. Flow cytometric analysis of androgen receptor expression in human prostate tumors and benign tissues. Clin Cancer Res 2000 6:1922–1930.

The three references marked with asterisks show the use of the propidium iodide staining method for the correlation of DNA content and cell cycle with the expression of nuclear and cytoplasmic markers in tumor cells.

5. Modified Click-iT™ Cell Proliferation Assay

Ronald M. Hamelik
rhamelik@med.miami.edu
Awtar Krishan
akrishan@med.miami.edu

Introduction

The following protocol is based on modifications made by Hamelik and Krishan (2009) to the protocol in the Invitrogen Click-iT™ Alexa Fluor® 488 EdU cell proliferation assay kit.

Reagents
Solutions not in the Invitrogen Kit

- 1% bovine serum albumin (BSA, Sigma) in phosphate buffered saline (PBS, pH 7.1-7.2).
- Deionized water or glass distilled water, 500 ml store at 2-6°C.
- Cell Lysis Solution: For the Hamelik & Krishan modified procedure; you will need 0.1% trisodium citrate dihydride (Sigma) containing 0.03% Nonidet P-40 (NP-40 detergent) in deionized water. Igepal CA-630 is an NP-40 equivalent available from Sigma-Aldrich, St. Louis, MO.
- Propidium iodide (Sigma), 4 µg/ml in PBS.
  Storage: Solutions should be stored in a refrigerator at 2-6°C.

Preparing the Stock Solutions

Allow all the stock solution vials in the Invitrogen kit to warm to room temperature before opening.

- EdU (Component A): to make a 10 mM solution, add 4 ml of DMSO (Component C) to component A and mix well.
- Alexa Fluor®488 Azide (Component B): for working solution add 260 µl DMSO (Component C) and mix well.
- Click-iT™ EdU reaction buffer (Component G):   DO NOT dilute the 10X stock solution at this time. Make the 1X reaction buffer fresh each time by mixing 1:10 with deionized water.
- Click-iT™ EdU buffer additive (Component I): To make a 10X stock solution of the Component I, add 4 ml of deionized water and mix until fully dissolved.
  Storage: Solutions should be stored in a refrigerator at 2-6°C and should be good for 6 months.

Method

1. Add 1 µl of the 10 mM EdU stock solution for each ml of culture media to a log-phase cell culture and incubate for 30-60 min at 37°C. We seed our suspension cultures at 0.5×10^6 cells 24 hours prior to doing our experiments so that they are at approximately 1×10^6 cells at the time of the experiment.

2. For suspension cultures, take a 2 ml aliquot (2×10^6 cells) out and collect cells by centrifugation for 5 minutes at 200 RCF and decant. For adherent cultures, remove the media, trypsinize the cells, resuspend in fresh media, count and pellet 2×10^6 cells by centrifugation for 5 minutes at 200 RCF and decant.
3. Resuspend the cell pellet in 0.5 ml of the cell lysis solution, vortex vigorously for 15 seconds.
4. Immediately add 2.5 ml of 1% BSA in PBS.
5. Centrifuge for 5 minutes at 4°C at 200 RCF and remove supernatant.
6. Prepare the Click-iT™ reaction cocktail according to the table below remembering that once prepared, the reaction mixture is only good for 15 minutes.

Reaction Component	Number of reactions		
	1	2	3
10X Click-iT™ Reaction buffer (G)	43.8 µl	87.5 µl	131.3 µl
Deionized water	393.8 µl	787.5 µl	1181.3 µl
CuSO$_4$ (H)	10 µl	20 µl	30 µl
Reaction Buffer Additive (I)	50 µl	100 µl	150 µl
Fluorescent Dye Azide (B)	2.5 µl	5.0 µl	7.5 µl
Total Volume	**500 µl**	**1000 µl**	**1500 µl**

7. Add 0.5 ml of the Click-iT™ reaction cocktail to each tube and mix well.
8. Incubate for 30 minutes at room temperature, **protected from light**.
9. Add 2.5 ml of 1% BSA in PBS. Centrifuge 5 minutes at 200 RCF and remove supernatant.
10. Resuspend the pellet in 1 ml of propidium iodide (4 ug/ml) solution and incubate for 30 minutes in the dark.

Set Up and Analysis

1. For analysis of the Alexa Fluor® 488 fluorescence, use 488 nm excitation with a green emission filter (530/30 nm or similar). Data should be normally collected on Y-axis in linear scale.
2. For cell cycle analysis using propidium iodide (PI), use 488 nm excitation with a red emission filter (610/20 nm or similar).
3. Set up a Forward vs. Side Scatter dot plot for gating out debris and a Peak vs. Integral (Height vs. Width) fluorescence histogram to gate out doublets (Figure not shown.)

Typical results are shown below in Figure 1.

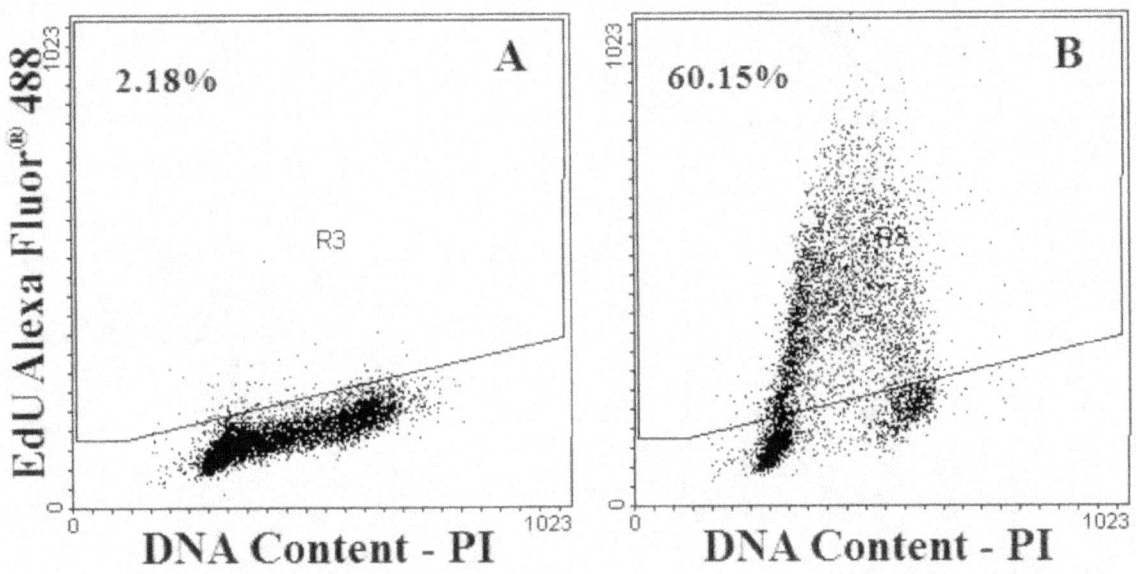

Figure 1. K562 triploid leukemia cells incubated with EdU for 45 minutes and then processed by the Hamelik-Krishan modified Click method. Cells prepared by Aysin Tulunay, University of Marmara Medical School, Istanbul, Turkey during the 2nd Turkish-US Flow Cytometry Workshop 2009

References
Hamelik R, Krishan A. Click-iT™ Assay with Improved DNA Distribution Histograms. Cytometry 2009 75A:862-865.

Krishan A, Hamelik R. Click-iT Proliferation Assay with Improved DNA Histograms. Current Protocols in Cytometry 2010 Unit 7.36:7.36.1-7.36.7.

6. Apoptotic Cell Analysis

Michael G. Ormerod
m.g.ormerod@btinternet.com

Introduction
Below are given three protocols for methods commonly used to enumerate apoptotic cells in cell cultures. When a cell becomes apoptotic, in the absence of scavenging cells, as would be found in vivo, the plasma membrane eventually ruptures and the cell contents degrade and are released. The rate at which this happens depends on the cell and the insult used to induce apoptosis. Consequently, these methods are recording cells in a transient population; their numbers depending on the rate of production balanced against the rate of cell loss.

Before using any of these methods, the presence of apoptotic cells should be confirmed by microscopy. Flow cytometry on its own should not be relied on to identify, without doubt, apoptotic cells. Most assays are capable of giving false positive and false negative results.

In adherent cell cultures, the apoptotic cells will usually detach from the culture dish surface. The culture medium should always be collected and checked for the presence of cells. It is advisable to record data separately from the detached and attached cells.

For reviews on flow cytometry and apoptosis, see publications by Darzynkiewicz et al. (1994), Ormerod (2001) and Ormerod (2008).

1. Measuring the 'sub-G_1 peak' in the DNA histogram
This method relies on the extensive degradation of DNA and nuclear fragmentation during the later stages of apoptosis. If cells are fixed in ethanol and subsequently rehydrated, some of the lower molecular weight DNA leaches out, lowering the DNA content. These cells along with nuclear fragments can be observed as a hypodiploid or 'sub-G_1' peak in a DNA histogram.

Reagents
- PI stock solution. 400 µg/ml propidium iodide (PI) (Sigma or Invitrogen) in water. Store at 4°C.
- RNase solution. Prepare just before use by dissolving 1 mg/ml RNase (Sigma) in phosphate buffered saline (PBS), pH. 7.5-8.0.

Method
1. Prepare a single cell suspension in 200 µl of phosphate buffered saline (PBS).
2. Add vigorously 2 ml of ice-cold 70% ethanol, 30% distilled water. Leave for at least 30 min on ice.
3. Harvest the cells by centrifugation. Resuspend in 400 µl of PBS.

4. Check the cells under a microscope. If they are clumped, pass through a 25 gauge syringe needle.
5. Add 50 µl of RNase and 50 µl of PI solutions. Incubate at 37°C for 30 min.
6. Analyse using an argon-ion laser tuned to 488 nm and measuring forward and side light scatter and red fluorescence using a linear amplifier (measuring area and either peak or width of the fluorescent signal). Display peak or width of the red signal against its area and set a region on the single cells to exclude doublets; use this region as a gate to display the DNA histogram.

A typical result is shown in Figure 1. Figure 2 shows data from an adherent cell culture.

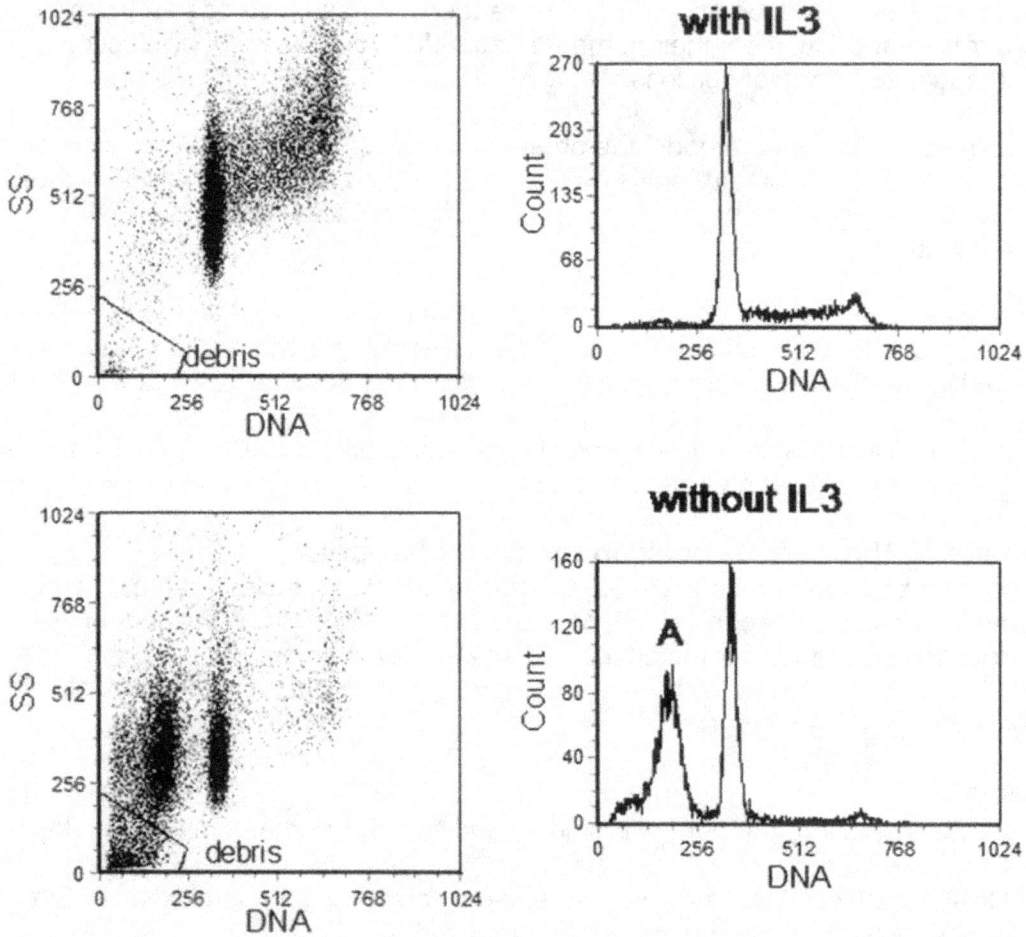

Figure 1. Murine haemopoietic cell line, BAF3, grown for 16 hours either with or without the growth factor, IL-3. Cells fixed in 70% ethanol and resuspended in PBS with PI. The DNA histograms have been gated to include only single cells (plot of DNA peak versus area not shown) and to exclude debris. 'A' marks the sub-G_1 peak (apoptotic cells). Figure reproduced, with permission, from Ormerod (2008).

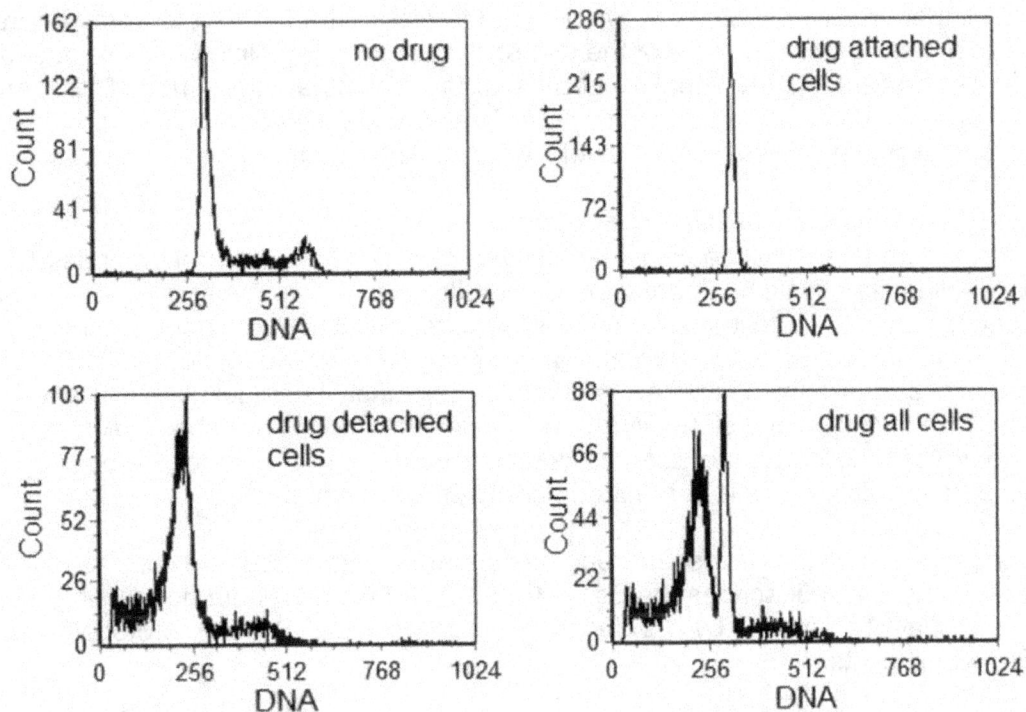

Figure 2. Ovarian carcinoma cell line, IGROV1, incubated with a kinase inhibitor for 24 h. The detached cells gave a clear sub-G_1 peak. Cells fixed in 70% ethanol. PI stain. Cells supplied by Bhaskar Bhattacharya, Institute of Cancer Research, UK. Figure reproduced, with permission, from Ormerod (2008).

Comments

- After fixation (step 2), the cells can be stored at -20°C.
- Sometimes a cytogram of side scatter versus DNA area can be used to gate on unwanted debris (Figure 1).
- The amount of DNA extracted from the cells, and hence the position of the apoptotic peak in the DNA histogram, depends on the type of cell under study. Sometimes the apoptotic cells may fail to give a 'sub-G_1' peak although an inspection of a cytogram of DNA peak versus DNA area may reveal the apoptotic cells. In such cases, after ethanol fixation, the cells should be resuspended in a phosphate-citrate buffer (0.2 M Na_2HPO_4, 4 mM citric acid, pH 7.8), a treatment which will extract more DNA from the apoptotic cells (Gong et al., 1994).
- Care should be taken in interpreting data from a DNA histogram. Apoptotic cells should be observed as a distinct peak. Necrotic cells, whose DNA is degraded randomly, will have a reduced DNA content and will be distributed across the same region of the histogram. If the instrument has been set to trigger on DNA/dye fluorescence (as is usually done when measuring a DNA histogram), an artificial 'sub-G_1' peak may be observed.

- A linear amplifier should always be used to record the DNA histogram. There are examples in the literature of the use of a logarithmic amplifier. Not infrequently, the so-called 'sub-G_1' peak will be found to contain particles whose DNA content is a few percent of that of cells in G_1. One has to question whether particles containing such small amounts of DNA should be counted as apoptotic cells.

2. TUNEL method for labelling apoptotic cells

Using the enzyme, terminal deoxynucleotidyl transferase (Tdt), the broken ends of DNA can be labelled directly using fluorescein-deoxyuridine triphosphate (dUTP) or indirectly using either biotin-dUTP followed by fluorescein-streptavidin, digoxygenin-dUTP followed by fluorescein-anti-digoxygenin or bromodUTP followed by fluorescein-anti-BrdUrd. The cells are fixed in paraformaldehyde to crosslink the cellular DNA and prevent the low molecular weight DNA from being extracted. When the DNA strand breaks have been labelled, the cells are counter-stained with PI so that the position in cell cycle from which the cells committed apoptosis can be observed.

One can purchase all the necessary reagents separately and label the cells as in the protocol below, which describes the BrdUTP method. I would recommend using one of the commercial kits supplied by a wide variety of companies. Suitable protocols are supplied with the kits.

Reagents
- Paraformaldehyde (Sigma). Make up a 1% solution in PBS. Adjust pH to 7.0. Either make up fresh or aliquot and store frozen.
- Tris-buffered saline (TBS): 150 mM NaCl in 50 mM Tris-HCl buffer, pH 7.6. Dissolve 6.06 g Tris (base), 1.39 g Tris-HCl, 8.8 g NaCl in distilled water and make up to 1 litre.
- Reaction buffer (x5): 1 M Na cacodylate, 125 mM Tris-HCl, pH 6.6, 1.25 mg/ml bovine serum albumen (BSA) (Cohn Fraction V, Sigma). Dissolve 1.96 g Tris-HCl, 16 g Na cacodylate and 0.125 g BSA in distilled water and make up to 100 ml. Store at 4°C.
- Enzyme mixture: Immediately before use, for each sample to be analysed, mix 20 Units terminal deoxynucleotidyl transferase (Tdt) (Pharmacia), 2 µl 2 mM BrdUTP (Sigma), 5 µl 25 mM $CoCl_2$, 10 µl reaction buffer and 35 µl distilled water.
- Rinsing buffer: PBS, 0.2% Triton X-100, 1% BSA. Add 100 µl Triton X-100 to 100 ml PBS, 1% BSA; warm to 37°C to dissolve Triton. Store at 4°C.
- Incubation buffer: 0.6 M NaCl, 60 mM Na citrate, 0.2% Triton X-100, 1% BSA. Dissolve 3.5 g NaCl, 1.76 g Na citrate, 1% BSA and 100 µl Triton X-100 in 100 ml distilled water. Warm to 37°C. Store at 4°C.
- Propidium iodide (PI) solution at 1 mg/ml (Sigma or Invitrogen) in distilled water. Store at 4°C.
- Ribonuclease (RNase) solution (Sigma). Dissolve 1 mg in 1 ml distilled water. Make up fresh just before use.

Method

1. Make a suspension of approximately 2×10^6 cells single cells.
2. Centrifuge at 300 g for 5 min at room temperature. Discard the supernatant and resuspend the pellet thoroughly in 100 µl PBS.
3. Add 1 ml ice-cold 1% paraformaldehyde solution. Leave on ice for 15 min.
4. Centrifuge as in Step 2. Resuspend the pellet in ice-cold PBS. Centrifuge and resuspend thoroughly in 200 µl PBS.
5. Add vigorously 2 ml ice-cold 70% ethanol. Leave at least 30 min on ice. The cells can be stored indefinitely at -20°C in this state.
6. Centrifuge cells at 400 g for 5 min at room temperature, discard supernatant, resuspend pellet in TBS. Centrifuge cells at 300 g for 4 min, resuspend cell pellet in 50 µl enzyme mixture. Incubate at 37°C for 1 h.
7. Centrifuge cells at 300 g for 4 min, resuspend cell pellet in rinsing buffer. Repeat this washing procedure and finally resuspend the cell pellet in 100 µl incubation buffer containing anti-BrdUrd-fluorescein antibody at 1mg/ml. Incubate on ice for 30 min.
8. Wash cells twice in rinsing buffer (as in Step 7) and resuspend final pellet in 430 µl PBS. Add 50 µl RNase solution, 20 µl PI solution. Incubate for 30 min at 37°C.
9. Analyse on the flow cytometer using blue light (argon-ion laser at 488 nm) and measuring forward and side light scatter and green and red fluorescence. Use doublet discrimination by measuring area and either peak or width of the red fluorescent signal. Measure the green fluorescence using a logarithmic amplifier and the red using a linear amplifier. Set the discriminator (threshold) to exclude debris and gate on the red fluorescent signal. Display peak or width of the red signal against its area and set a region on the single cells; use this region as a gate to display a cytogram of green fluorescence (fluorescein; strand breaks) versus red fluorescence (PI; DNA). The apoptotic cells are positive for green fluorescence.

Typical results are shown in Figure 3.

Comments

- After fixation, the cells can be stored indefinitely at -20°C.
- If there a few apoptotic cells present, particularly in clinical specimens, it is helpful to sort the positive events onto a slide for confirmation by light microscopy (Dowsett et el., 1998).

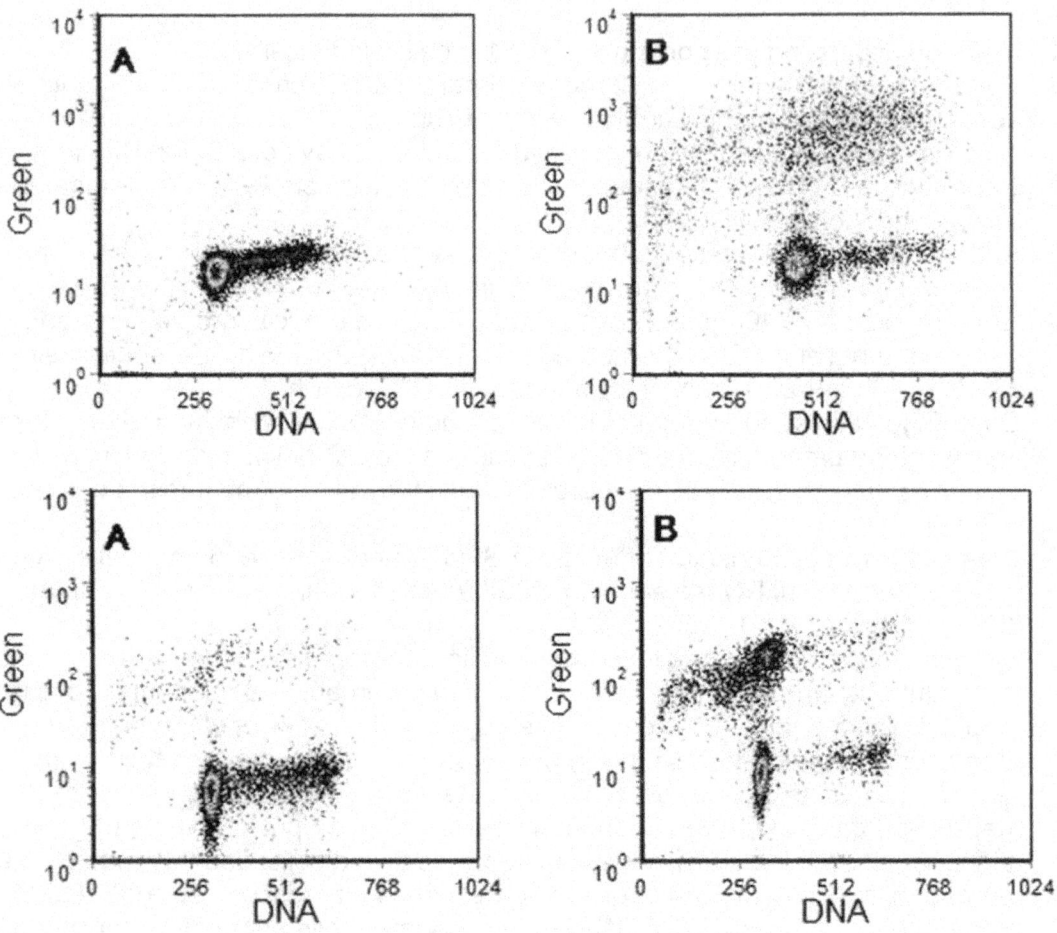

Figure 3. Upper panel. HL60 (human promyelocytic cell line) cells. A, control cells; B, incubated for 4 hours with camptothecin. TUNEL assay. Cells supplied by Phoenix Flow Systems. Lower panel. Cells as described in Figure 1. A, incubated with and B without IL-3. Labelled using the TUNEL assay. Cells prepared by Simone Detre, Academic Dept. of Biochemistry, Royal Marsden Hospital, London.

3. Staining with Annexin V

Annexin V is a protein with high affinity for negatively charged phospholipids, such as phosphatidyl serine (PS), in the presence of Ca^{2+} ions. Unfixed cells are incubated with Annexin V conjugated to a suitable fluorochrome. Several manufacturers supply Annexin V conjugated to a variety of fluorochromes. It is customary to add PI to distinguish cells that have lost integrity of the plasma membrane, although, if Annexin V is to be combined with immunochemical labels, PI could be omitted.

Reagents
- 10 mM Herpes/NaOH pH 7.4, 140 mM NaCl, 2.5 mM CaCl2
- FITC-Annexin V
- PI stock solution. 400 µg/ml propidium iodide (PI) (Sigma or Invitrogen) in water. Store at 4°C

Method
1. Prepare a single cell suspension at $1-2 \times 10^6$ cells/ml. Wash cells in PBS.
2. Centrifuge and resuspend in Hepes/NaOH buffer.
3. Add FITC-Annexin V to a final concentration of 1 µg/ml.
4. Incubate 10 min in the dark at room temperature.
5. Add PI to a final concentration of 2 µg/ml. Incubate a further 5 min.
6. Analyse recording right angle and forward light scatter, log green (520 nm) and log red fluorescence (>650 nm). (Use the fluorescein filter for green fluorescence and a deep red filter for red).

Figure 4 shows a typical result. If a gate is set on light scatter to exclude clumps of cells and debris, care must be taken not to exclude any apoptotic cells, which have changed light scatter compared to the viable cells.

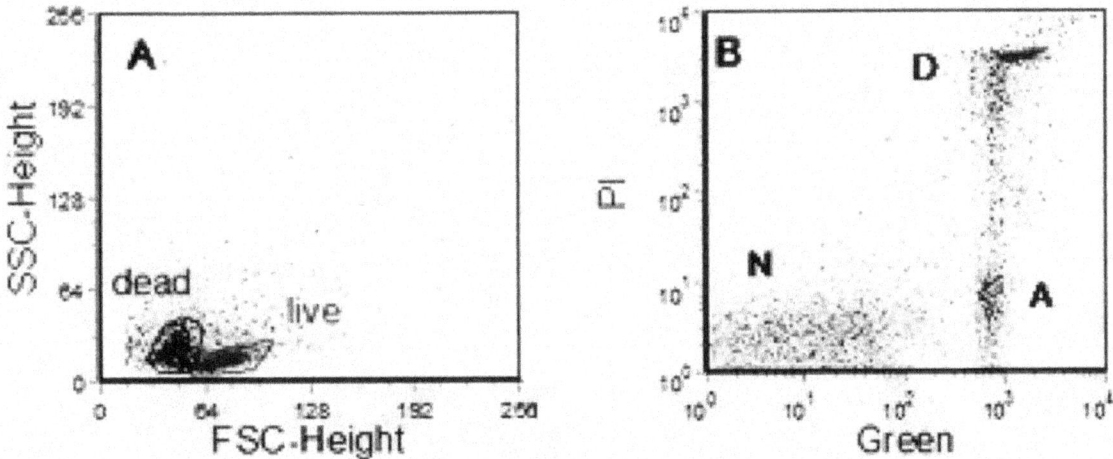

Figure 4. Human chronic lymphocytic leukaemia cells incubated in culture with chlorambucil followed by Annexin V-FITC. PI added. A. Light scatter. B. PI versus Annexin V. The labels indicate normal (N), apoptotic (A) and 'necrotic' (D) cells. Data supplied by Chris Pepper, Department of Haematology, Llandough Hospital, Penarth, Wales. Figure reproduced, with permission, from Ormerod (2008).

References
Darzynkiewicz Z, Li X, Gong J. Assays of cell viability: discrimination of cells dying by apoptosis. Methods in Cell Biology 1994 41:15-38.

Dowsett M, Detre S, Ormerod MG, Ellis PA, Mainwaring PN, Titley JC, Smith IE. Analysis and sorting of apoptotic cells from fine needle aspirates of excised primary breast carcinomas. 1998 Cytometry 32:291-300.

Gong J, Traganos F, Darzynkiewicz Z. A selective procedure for DNA extraction from apoptotic cells applicable for gel electrophoresis and flow cytometry. Anal Biochem 1994 218:14-19.

Ormerod MG. Flow Cytometry – a basic introduction. IBSN 978-0-9559812-0-3. Self published. 2008 Available from www.denovosoftware.com.

Ormerod MG. Using flow cytometry to follow the apoptotic cascade. Redox Report 2001 6:275-287.

Van Engeland M, Nieland LJW, Ramaekerss FCS, Schutte B, Reutelingsperger CPM. Annexin V-affinity assay: a review on an apoptosis detection system based on phosphatidylserine exposure. Cytometry 1998 31:1-9.

7. Neutrophil Apoptosis Assay

Sachin Kumar
sach_bio@yahoo.co.in
Madhu Dikshit
madhu_dikshit@cdri.res.in

Introduction
Neutrophils (PMNs) are short-lived cells and at inflammatory sites they undergo apoptosis and are subsequently engulfed by macrophages (Savill et al., 2002). In contrast to apoptosis in other cell types, substantial cell shrinkage, membrane blebbing, and formation of apoptotic bodies are not prominent features of apoptotic neutrophils.

Apoptosis involves processing and activation of pro-apoptotic Bcl-2 family member proteins (e.g., Bax), release of cytochrome c and activation of caspases 3. During apoptosis, phosphatidylserine in the cell membrane is translocated to the outer surface of the plasma membrane (Savill et al., 2002). In apoptosis, endonuclease cleave internucleosomal DNA into monomers/oligomers or nucleosomal DNA into 200 base pairs and multiples, which results in oligonucleosomal DNA fragmentation.

Neutrophil apoptosis may be analyzed by morphological changes, plasma membrane alterations, mitochondrial membrane potential loss, caspase activation, and DNA fragmentation (Luo and Loison, 2008). Flow cytometry is a rapid and reliable method for identification and analysis of the apoptotic cells and to identify the mechanisms involved.

1. Annexin V-PI Apoptosis Assay
Phospholipids are asymmetrically distributed between the inner and outer leaflets of the plasma membrane. Early stage of apoptosis is characterized by translocation of phosphatidylserine from the inner to the outer layer of the plasma membrane (Vermes et al., 1995). Annexin V is a phospholipid-binding protein with high affinity for negatively charged phosphatidylserine in the presence of Ca^{++} ions. Thus fluorescent labelled Annexin V and propidium iodide can be used for rapid identification of viable, apoptotic and necrotic cells by flow cytometry.

Reagents
- Fluorescein-conjugated annexin V (Annexin V FITC kit, BD biosciences)
- Annexin V binding buffer (10 mM Hepes-NaOH,pH 7.5; 140 mM NaCl; 2.5 mM $CaCl_2$)
- Propidium iodide (PI): 1 mg/ml in distilled water; store at 4^0C in the dark

Method
1) Incubate the freshly isolated peripheral blood neutrophils at 1×10^6 cells/ml in RPMI medium containing 10% FBS with or without apoptosis-modifying agents in 48 well culture plate.

2) Incubate the plate at 37°C in a 5% CO_2 incubator for the desired length of time.
3) Resuspend cells by gentle pipetting and centrifuge at 200 g for 2 min at room temperature. Discard the supernatants.
4) Resuspend the cells in 100 µl of Annexin V binding buffer (10 mM Hepes-NaOH,pH 7.5; 140 mM NaCl; 2.5 mM $CaCl_2$).
5) Incubate for 15 minutes in the dark at room temperature in presence of 5 µl of annexin V (1 µM), no wash is required.
6) Co-label membrane permeable necrotic or late apoptotic cells by staining with the membrane impermeant nuclear dye propidium iodide (PI; 1 µg/mL) 5 min prior to running the sample.
7) Following incubation, add 400 µl of binding buffer to each sample.
8) Analyze cells on the flow cytometer, using 488-nm excitation. Set gates based on light scatter. Collect green annexin V fluorescence at 530 nm (FL-1) and red PI fluorescence above 600 nm (FL-2 channel) following appropriate compensation.
9) Live cells are annexin V-negative and PI-negative; apoptotic cells are annexin V positive and PI-negative; necrotic cells are both annexin V-positive and PI-positive.

A typical result is shown in Figure 1.

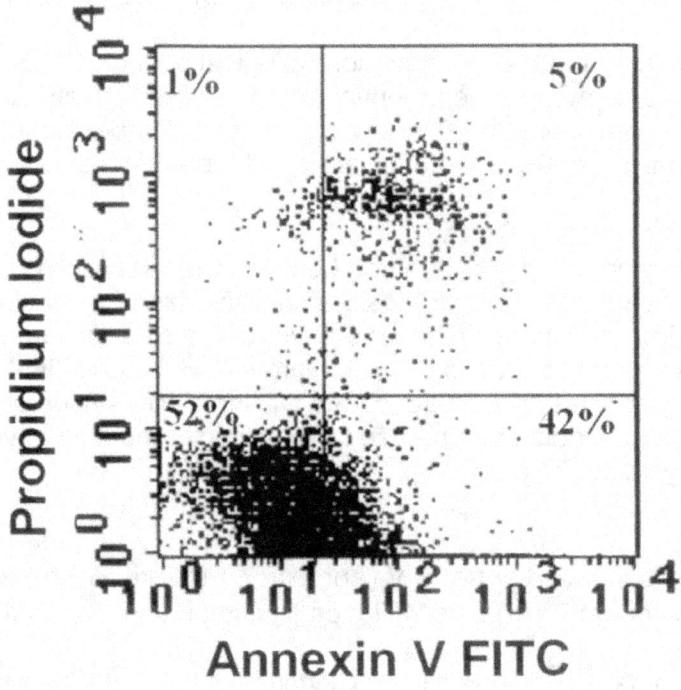

Figure 1. Annexin V & PI staining of neutrophils treated with NO donor sodium nitroprusside (SNP- 1 mM) for 1 h. Live cells representing annexin V-negative and PI-negative population (LL 52%); apoptotic cells are annexin V positive and PI-negative (LR 42%); necrotic cells are both annexin V-positive and PI-positive (UL 1%+ UR 5%).

Comments
- Analyze the samples by flow cytometry within one hour as apoptotic cells increase with time.
- Neutrophil isolation method and culture can profoundly influence apoptosis and the response to stimuli.
- Contamination with other cells like eosinophils and monocytes can influence neutrophils apoptosis.

2. Mitochondrial Membrane Potential Assay

Changes in neutrophil mitochondrial membrane potential may be measured using JC-1 (5,5',6,6'-tetrachloro-1,1',3,3'-tetra-ethyl-benzimidazocarbo-cyaniniodide), a cationic dye which exhibits membrane potential-dependent accumulation in mitochondria indicated by a fluorescence emission shift from green (525 nm) to red (590 nm). In viable cells with intact mitochondrial membrane potential, JC-1 is able to enter mitochondria and emit red fluorescence. When mitochondrial membrane potential collapses during apoptosis, JC-1 cannot accumulate within the mitochondria and remains in the cytoplasm in a green fluorescent monomeric form (Kumar et al., 2010). Apoptosis associated mitochondrial depolarisation is indicated by a decrease in the red/green fluorescence intensity ratio.

Reagents
- JC-1 (1 mM) stock solution dissolved in DMSO

Method
1. Incubate the freshly isolated peripheral blood neutrophils at 1×10^6 cells/mL in RPMI containing 10% FBS ± apoptosis-modifying agents in 48 well culture plates.
2. Incubate the plate at 37°C in a 5% CO2 incubator for the desired length of time.
3. Centrifuge at 200 g for 5 min and discard the supernatants.
4. Gently resuspend cell pellet in 1 ml of freshly prepared 2.5 µM JC-1 solution in PBS and incubate at 37°C for 15 min.
5. Analyze e on a flow cytometer, using 488-nm excitation. Collect green fluorescence at 530 ± 20 nm (FL1 Channel) and red fluorescence at 570 ± 20 nm (FL2 Channel).
6. Calculate the ratio of red vs. green fluorescence.

A typical result is shown in Figure 2.

Comments
- Addition of JC-1 to PBS causes precipitation which may induce neutrophils clumping and apoptosis, while addition of PBS to the small amount of JC-1(from stock) does not lead to precipitation.
- The collapse of mitochondrial membrane potential may not be a prerequisite for release of cytochrome c, AIF, and other apoptotic events, one should be cautious in interpreting the data.

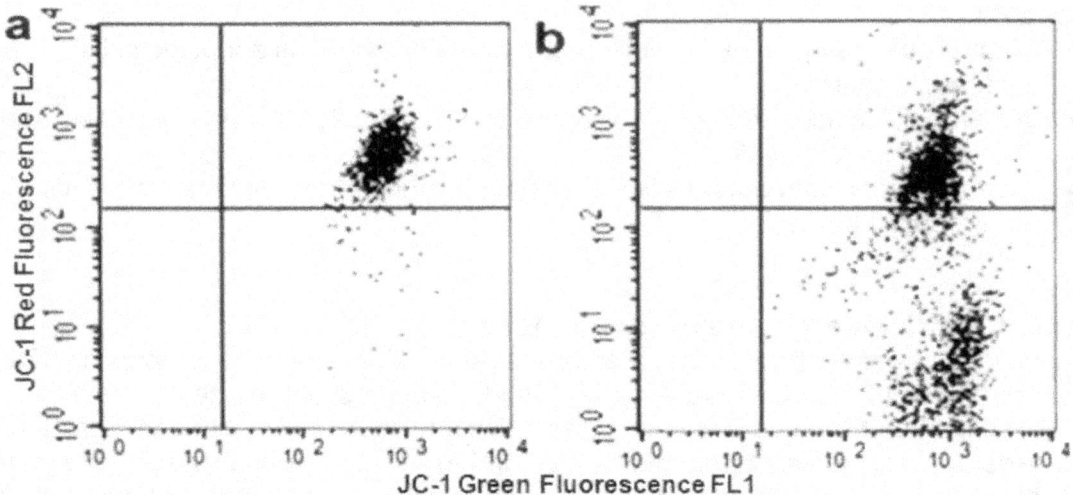

Figure 2. Mitochondrial membrane potential loss in neutrophils as measured by JC-1. (a) control neutrophils (b) Cells treated with NO donor sodium nitroprusside (SNP- 1 mM) for 30 min. Cells in lower right quadrant show the loss of mitochondrial membrane potential.

3. TdT-Mediated dUTP Nick End Labeling, TUNEL Assay

In the later stage of apoptosis, DNA fragmentation takes place due to the activation of endonucleases. Terminal deoxynucleotidyl transferase (TdT) enzyme catalyzes a template-independent addition of fluorescein-deoxyuridine--triphosphate (FITC-dUTP) to the 3'-hydroxyl (OH) termini of double- and single-stranded DNA (Kumar et al., 2010). After incorporation, these sites can be identified by flow cytometry.

Reagents

- APO-Direct^{TM} Kit: (BD Biosciences) containing various components
- 70% ethanol: ice cold
- Propidium Iodide/RNase Staining Buffer: 5 µg/ml PI, 200 µg/ml RNase in PBS.
- Rinsing buffer: 0.1% (v/v) Triton X-100 and 5 mg/ml BSA in PBS, pH 7.4
- FITC-dUTP: 0.25 nMol/reaction; contains 0.05% sodium azide.
- TdT Enzyme: (S.A.= 100,000 U/mg) 200 µg/ml in 50% (v/v) glycerol solution **Do not freeze at –20°C.**
- Negative and positive control cells: contains 70% (v/v) ethanol.
- TdT staining buffer: (5x) 1 M potassium or sodium cacodylate,125 mM Tris-Cl,1.25 mg/ml BSA pH 6.6.

Method

1. Incubate the freshly isolated peripheral blood neutrophils at 1×10^6 cells/ml in RPMI medium containing 10% FBS ± apoptosis-modifying agents in 48 well culture plates.
2. Incubate the plate at 37°C in a 5% CO_2 incubator for the desired length of time.
3. Centrifuge neutrophil suspension at 300 g for 5 min at 4°C. Discard the supernatant.
4. Fix the cells with 1% paraformaldehyde for 20 min on ice and wash twice with PBS 300 g for 5 min at room temperature.
5. Resuspend the pellet in chilled 70% ethanol overnight.
6. Wash the experimental and positive cells (introduced DNA strand breaks with DNase I) twice with PBS at 300 g for 5 min.
7. Incubate the cells in 50 µl reaction buffer (staining buffer 5x (10 µl), TdT enzyme (0.75 µl), FITC dUTP (8.00 µl), 32 µl water) for 60 min at 37^0C.
8. Stain the cells with PI/RNase solution for 30 min.
9. Analyze samples on a flow cytometer within 2 h of labeling.

Comments

- Cells in 70% ethanol can be stored at –20°C for several days prior to use.
- It is necessary to include positive and negative controls as DNA strand break labelling is rather complex and involves many steps.
- The method described is easy and uses direct labeling of FITC-dUTP. DNA nick labeling can be done indirectly using Br-dUTP followed by FITC-anti-BrdU monoclonal antibody.
- The flow cytometry method offers a unique opportunity to analyze the cell cycle stage of cells undergoing apoptosis.

References

Kumar S, Barthwal MK, Dikshit M. Cdk2 nitrosylation and loss of mitochondrial potential mediate NO-dependent biphasic effect on HL-60 cell cycle. Free Radic Biol Med 2010 48:851-861.

Luo HR, Loison F. Constitutive neutrophil apoptosis: mechanisms and regulation. Am J Hematol 2008 83:288-295.

Savill J, Dransfield I, Gregory C, Haslett C. A blast from the past: clearance of apoptotic cells regulates immune responses. Nat Rev Immunol 2002 2:965-975.

Vermes I, Haanen C, Steffens-Nakken H, Reutelingsperger C. A novel assay for apoptosis. Flow cytometric detection of phosphatidylserine expression on early apoptotic cells using fluorescein labelled Annexin V. J Immunol Methods 1995 184:39-51.

8. Drug Retention and Efflux Assay

Ronald M. Hamelik
rhamelik@med.miami.edu
Awtar Krishan
akrishan@med.miami.edu

Introduction
This protocol is used to analyze fluorescent drug retention and efflux in drug resistant cells with MDR phenotype and p-glycoprotein efflux pump that can be blocked by Verapamil. The procedure also allows for the identification of dead cells in the sample.

Reagents
Stock Solutions
- Rhodamine 123 (R123) (Sigma): 100 µg/ml in distilled water.
- Verapamil (Sigma): 1 mg/ml in phosphate buffered saline (PBS).
- Daunomycin: 1 mg/ml in PBS.

Working Solutions
- Rhodamine 123: 1 µg/ml in PBS, 1:100 dilution of the stock solution.
- Daunomycin: 1-10 µg/m in PBS, 1:1000 to 1:100 dilution of the stock solution.
- Verapamil: 100 µg/ml in PBS, 1:10 dilution of the stock solution.
- Propidium Iodide (Sigma): 50 µg/ml in PBS.
  Solutions should be stored in a refrigerator at 2–6˚C.

Cellular Drug Retention and Efflux with PI to Identify Dead Cells:
1. To 1.6 ml of cells @ 10^6/ml, add 200 µl of the working Rhodamine 123 or Daunomycin solution.
2. Add 200 µl of the working Propidium Iodide solution.
3. Split the sample (1.0 ml) into two test tubes marked Control and Blocked.
4. To the Control tube add 100 µl of PBS and to the Blocked tube add 100 µl of the working Verapamil solution.
5. Incubate for 10 minutes in a 37°C water bath.
6. Analyze on flow cytometer as described below.

Set Up and Analysis
- Rhodamine 123 and Daunomycin use 488 nm excitation and are measured with a yellow band pass emission filter (575/±30 nm).
- The flow cytometer workspace screen should display following plots:
  a. A dual parameter Side Scatter vs. Forward Scatter dot plot (SS vs. FS).
  b. A single parameter FL2 Log (575 BP channel) histogram.
  c. A dual parameter FL2 Log vs. Forward Scatter dot plot.
  d. Acquisition should stop at 10,000 events.

Typical Results

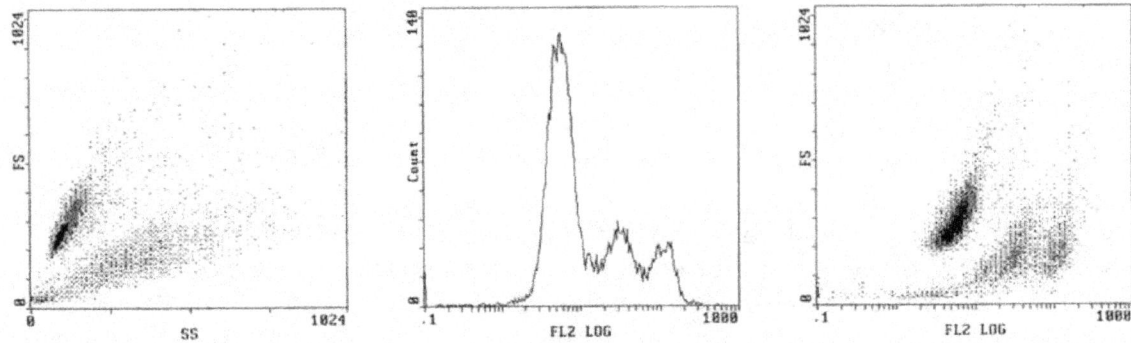

Figure 1 shows data from a suspension culture of P388/R84 cells which contained cells with drug resistance (approximately 84 fold resistant to doxorubicin than the parental P388 cells), cells which had lost their drug efflux and dead cells. Cells were incubated with rhodamine 123 and propidium iodide. The dot plot of FS vs. SS on the left shows the presence of two distinct scatter populations. In the histogram of rhodamine retention (middle) and dot plot of FS vs. rhodamine retention (right), three distinct populations of cells can be seen: resistant cells with low drug retention (on the left), cells lacking efflux with mid level of drug retention (in the center) and dead cells stained with PI with high fluorescence (on the right).

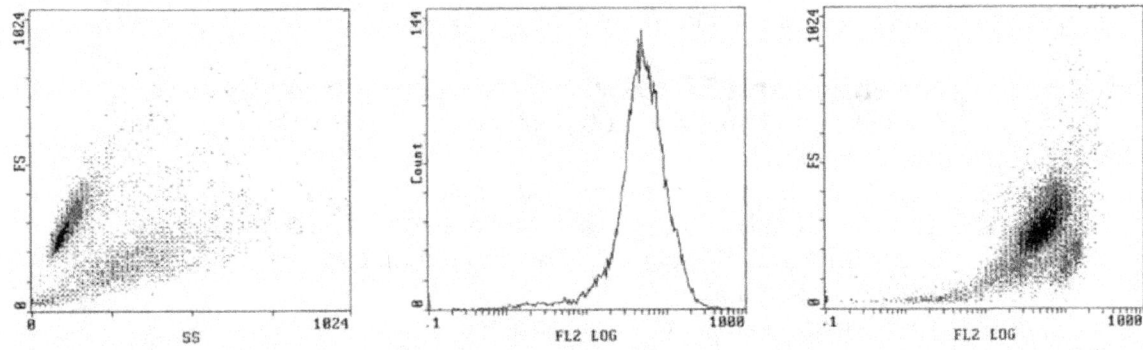

Figure 2 shows data from the same P388/R84 cells which were incubated with rhodamine 123, propidium iodide and the efflux blocker Verapamil. The middle histogram shows a large single population as the drug efflux was blocked in the resistant cells and they become as bright as the non-resistant cells and overlap the dead cells. In the dot plot on the right, two distinct populations are seen: 1. Drug resistant cells with blocked efflux pumps and cells lacking efflux pumps, both with a high retention of rhodamine 123. 2. Dead cells with broken cell membranes whose DNA has been stained with propidium iodide which makes them even brighter than the drug retaining cells.

References

Krishan A. Monitoring of cellular resistance to cancer chemotherapy: drug retention and efflux. Methods Cell Biol 2001 164:193-209.

Krishan A, Ganapathi R. Laser flow cytometric studies on the intracellular fluorescence of anthracyclines. Cancer Res 1980 40:3895-3900.

Krishan A, Hamelik, R., Flow cytometric analysis of drug transport and efflux in stem cells. Chapter 5. In: Application of flow cytometry in stem cell research and tissue regeneration. Editors: Awtar Krishan, H. Krishnamurthy and S. Totey. Pages 61-74, 2010, Wiley Blackwell, NY.

Krishan A, Sauerteig A, Wellham L. Flow cytometric studies on modulation of cellular adriamycin retention by phenothiazines. Cancer Res 1985 45:1046-1051.

Krishan A, Sridhar KS, Davila E, Vogel C, Sternheim W. Patterns of anthracycline retention modulation in human tumor cells. Cytometry 1987 8:306-314.

9. Neutrophil Phagocytosis Assay

Sachin Kumar
sach_bio@yahoo.co.in
Madhu Dikshit
madhu_dikshit@cdri.res.in

Introduction

Circulating neutrophils kill the invading pathogens by their ability to recognize and phagocytose invading bacteria which are killed by generation of highly reactive oxygen species and release of microbicidal proteases (Lee et al., 2003; Segal, 2005).

Although light microscopy can be used for direct assessment of bacterial phagocytosis by neutrophils, lack of high resolution makes it difficult to count small particles such as bacteria by this method. Electron microscopy can overcomes this problem but it is difficult to routinely use it due to the time required for sample preparation and analysis. Flow cytometry by analyzing fluorescent labelled particles (bacteria or beads) ingested by neutrophils has the advantage due to its high speed and sensitivity which can also easily distinguish between intracellular and extracellular cell membrane attached bacteria (Hampton and Winterbourn, 1999)

Reagents

- Luria-Bertani (LB) broth: (10 g Bacto tryptone, 5 g Bacto yeast extract, 10 g NaCl, H_2O to 1 liter; pH = 7.0, autoclaved. (LB agar; LB broth containing 15 g Bacto agar/liter).
- Phosphate buffer saline (PBS); 137 mM NaCl, 2.7 mM KCl, 10 mM Na_2HPO_4, 1.8 mM KH_2PO_4 (pH= 7.4).
- Trypan blue: 2 mg/ml in PBS.
- Fluoroscein Isothiocyanate (FITC): 5 mg/ml in PBS (pH 7.4).
- Hanks' Balanced Salt Solution (HBSS): NaCl 138 mM, KCl 2.7 mM, Na_2HPO_4 8.1 mM, KH_2PO_4 1.5 mM, Glucose 10 mM, pH 7.4 without Ca^{2+} or Mg^{2+}.
- HBSS containing Ca^{2+} Mg^{2+}; add 1 mM $CaCl_2$ and 1 mM $MgCl_2$.
- Neutrophils and bacteria (E.coli)

Preparation of Bacteria

1) Inoculate 5 ml of sterile Luria-Bertani (LB) broth with a single colony of *E.coli* grown on LB agar plates. Grow in a shaking incubator for 18 h at 37°C.
2) Take a 1 ml sample from the overnight culture and centrifuge at 10,000 *g* for 5 min at room temperature to pellet the bacteria. Wash the pellet twice by resuspending in PBS and centrifuging.
3) Calculate the concentration of bacteria in the sample by measuring the optical density at 550 nm and relating to a previously established standard curve of optical density vs. colony-forming units (CFU).

4) Bacteria are heat inactivated by suspending 1×10^8 CFU in 1 ml of PBS at 60°C for 30 minutes in a water bath.
5) Subsequently *E. coli* are labeled with FITC (50 μg/ml) in PBS (pH 7.4) in the dark at room temperature for 1 h.
6) Wash the FITC Labeled bacteria twice with PBS at 10,000 g for 5 min at 4°C. Keep the FITC labeled bacteria on ice until use.

Method

1. Incubate the freshly isolated peripheral blood neutrophils at 1×10^6 cells/ml in HBSS containing Ca^{2+} and Mg^{2+} in presence of FITC labeled bacteria at 1:50 for 30 minutes with or without phagocytosis-modifying agents.
2. Analyze cells on the flow cytometer, using 488 nm excitation. Set neutrophil gates based on light scatter. Collect green FITC fluorescence at 530 nm (FL-1 channel).
3. Add 40 μl of the Trypan blue stock solution and wait 5 minutes to differentiate between phagocytosed and adherent bacteria.
4. Re-analyze the sample on the flow cytometer.

Comments

i. A potential problem arises if viable bacteria divide during the assay, since daughter cells will contain less fluorescent label. Always use heat killed or fixed bacteria if bacterial killing efficiency of the neutrophils is not to be analyzed.
ii. Trypan blue quenches the fluorescence of adherent bacterial population and helps in discrimination between attachment and internalization.

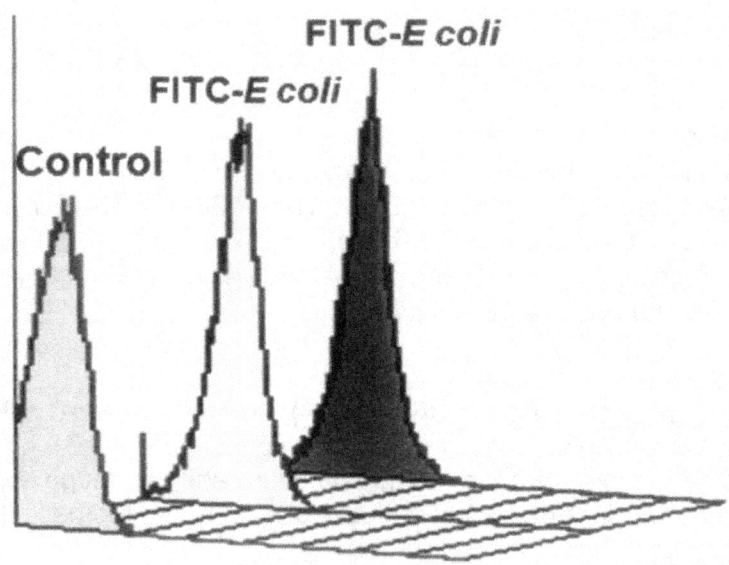

Figure 1. Histogram overlay of neutrophils incubated with FITC labeled bacteria. Cells were incubated with FITC labeled bacteria (E. coli) in PMN: bacteria ratio of 1:30 (middle histogram) and 1:50 (right panel).

References

Hampton MB, Winterbourn CC. Methods for quantifying phagocytosis and bacterial killing by human neutrophils. J Immunol Methods 1999 232:15-22.

Lee WL, Harrison RE, Grinstein S. Phagocytosis by neutrophils. Microbes Infect 2003 5:1299-1306.

Segal AW. How neutrophils kill microbes. Annu Rev Immunol 2005 23:197-223.

10. Respiratory Burst / Free Radical Assay

Sachin Kumar
sach_bio@yahoo.co.in
Madhu Dikshit
madhu_dikshit@cdri.res.in

Introduction

Circulating neutrophils kill the invading pathogens by phagocytosis and generation of highly reactive oxygen species and release of microbicidal proteases. Respiratory burst following phagocytosis by neutrophils is due to the activation of NADPH oxidase, a multi-subunit enzymatic complex (Babior et al., 2002). Upon stimulation, cytoplasmic proteins, $p47^{Phox}$, $p67^{Phox}$, and a Rac-related GTP protein translocate to the plasma membrane and bind to the sites on a unique b-type hemoprotein, Cytochrome b_{558} which binds FAD and NADPH that results in a flow of electrons to the terminal acceptor Cytochrome b_{558} (Babior et al., 2002; Segal, 2005). Transfer of an electron from the Cytochrome to oxygen yields superoxide O_2^- which forms additional toxic oxygen species, in particular H_2O_2, by spontaneous dismutation which oxidizes halides, in particular Cl^-, to form hypohalous acid, e.g. HOCl. The production of superoxide initiates a series of reactions to generate various reactive oxygen species, which result in microbial killing.

Reactive oxygen species (ROS) and reactive nitrogen species (RNS) formation in PMNs has been detected in our laboratory by using luminescence (luminol, lucigenin), fluorescence (DCF, DHE & DHR), absorbance (NBT, cytochrome c reduction) or electron paramagnetic resonance (EPR; by utilizing spin traps, DMPO/DEPMPO) (Dikshit et. al., 2002), Sharma, et al., 2004 and Tarpey et al., 2004). In the present protocol, we have used flow cytometry for detection of fluorochromes which indicate the presence of free radical generation by change in fluorescence emissions.

Reagents

- Hanks' Balanced Salt Solution (HBSS): NaCl 138mM, KCl 2.7mM, Na_2HPO_4 8.1mM, KH_2PO_4 1.5mM, Glucose 10 mM, pH 7.4 without Ca^{2+} or Mg^{2+}.
- HBSS containing Ca^{2+} Mg^{2+}: add 1mM $CaCl_2$ and 1mM $MgCl_2$.
- 2′,7′-dichlorofluorescin diacetate (DCFH-DA): MW 487; prepare 10 mM in 100% ethanol and keep on ice, protected from light.
- Dihydroethidine (DHE): MW 315; prepare fresh 10 mM in DMSO and keep on ice, protected from light.
- Dihydrorhodamine 123 (DHR): MW 346; prepare fresh 10 mM in DMSO. Keep on ice, protected from light.
- Formyl-methionyl-leucyl-phenylalanine (fMLP): MW 436.7; prepare 10 mM stock solution in DMSO and store in 10 µl aliquots ≤12 months at −20°C.
- Phorbol myristate acetate (PMA): MW 616.84; prepare 1 mM stock solution in DMSO and store in 10 µl aliquots at −20°C.

Method

1) Load the PMNs (2×10^6 cells/ml) with dye (DCF-DA at 10 μM or DHR-123 at 10 μM or DHE at 10 μM) in HBSS with Ca^{2+} Mg^{2+} by incubating them for 10 min in a 37°C water bath.

2) Add PMA (50 nM) or fMLP (1 μM) or other stimulator.

3) Incubate for 30 min in a 37°C water bath.

4) Acquire the sample after gating on the neutrophil population (using FSC/SSC or by using neutrophil specific antibodies) for free radical generation by acquiring a minimum of 10,000 cells in the FL1 channel (530±20 nm) for DCF, DHR and DAF or in the FL2 channel (570±20 nm) for DHE.

5) Calculate the free radical generation as mean fluorescence or mean stimulation index (MSI), a ratio of mean fluorescence of the stimulated vs. unstimulated cells.

6) Incubate the cells with propidium iodide (5 μg/ml) in another set of tubes to explore the effect of interventions on cell viability, acquire minimum 10,000 cells.

Comments

- 2′,7′-Dichlorodihydrofluorescein diacetate (DCDHF-DA) diffuses in to the cells where intracellular esterases hydrolyze the acetate group, forming 2′, 7′-dichlorofluorescin (DCHF), a non-fluorescent, polar substance to get trapped within the cells. During the intracellular production of ROS, DCHF is oxidized to highly fluorescent, 2′, 7′-dichlorofluorescein (DCF). DCHF reacts with most of reactive species including hydrogen peroxide and peroxynitrite hypochlorous acid, hydroxyl radical and nitryl chloride (NO_2Cl), and is most commonly used probe for oxidative stress.

- The oxidative fluorescent probe dihydroethidium (or hydroethidine, DHE) allows detection of O_2^-· generation. DHE is a lipophilic cell-permeable dye that can be rapidly oxidized to ethidium bromide by O_2^-·, and emits a bright red fluorescence.

- Dihydrorhodamine 123 (DHR-123) is an uncharged, nonfluorescent mitochondrion-avid dye. During the cellular production of free radicals DHR-123 is oxidized to rhodamine 123 and emits a green fluorescent signal. DHR-123 reacts with hydrogen peroxide and also reacts with peroxynitrite. 4, 5-Diaminofluorescein (DAF) is used to measure low levels of nitric oxide (NO), DAF is nonfluorescent until it reacts with NO to produce a green fluorescent product (DAF-2T).

- Cell loading of dyes generally takes 10-15 min in case of neutrophils, while other cells like endothelial cells takes ~ 45 min. The final concentration of vehicle (ethanol/DMSO) should be below the recommended 1% maximum. DMSO should be used as a vehicle control with each experiment.

- To determine the effect of various inhibitors/interventions, incubate the cells at 37°C for 15-30 min, before loading with the dye.

- For a time kinetic study, run the tube on the flow cytometer as soon as stimulator is added. Proceed to individually stimulate and run each experimental tube, and run control tubes without addition of stimulator. It is necessary to ensure identical conditions and timing for each sample.

- Mean stimulation index (MSI) is measured to prevent the day to day variation in fluorescence and report the data as fold change against control of each day.

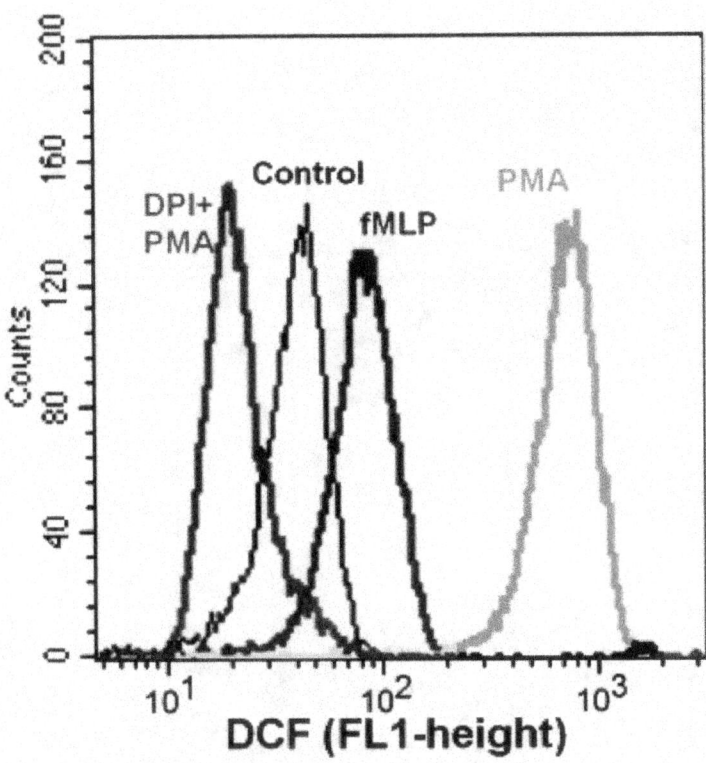

Figure 1. Overlay histograms showing free radical generation in neutrophils after stimulation with fMLP (1 µM) and PMA (50 nM), which was measured as DCF fluorescence. PMA stimulated ROS was significantly inhibited in cells pretreated with NADPH oxidase inhibitor (DPI-10 µM).

References

Babior BM, Lambeth JD, Nauseef W. The neutrophil NADPH oxidase. Arch Biochem Biophys 2002 397:342-344.

Dikshit M, Sharma P. Nitric oxide mediated modulation of free radical generation response in the rat polymorphonuclear leukocytes: a flowcytometric study. Methods Cell Sci 2002 24:69-76.

Segal AW. How neutrophils kill microbes. Annu Rev Immunol 2005 23:197-223.

Sharma P, Raghavan SA, Saini R, Dikshit M. Ascorbate-mediated enhancement of reactive oxygen species generation from polymorphonuclear leukocytes: modulatory effect of nitric oxide. J Leukoc Biol 2004 75:1070-1078.

Tarpey MM, Wink DA, Grisham MB. Methods for detection of reactive metabolites of oxygen and nitrogen: in vitro and in vivo considerations. Am J Physiol Regul Integr Comp Physiol 2004 286:R431-444.

11. Paroxysmal Nocturnal Haemoglobinuria Assay

Manisha Madkaikar
madkaikarm@icmr.org.in
Maya Gupta
Maya_rk@rediffmail.com

Introduction

Paroxysmal nocturnal haemoglobinuria (PNH) is an acquired stem cell disorder that results from the nonmalignant clonal expansion of hematopoietic stem cells harboring somatic mutations in an X-linked gene, *PIG-A*. Progeny of affected stem cells are deficient in glycosyl phosphatidylinositol–anchored proteins (GPI-AP). It is clinically characterized by intravascular haemolysis, thrombosis and varying degree of cytopenias. PNH is classified depending on the presence or absence of underlying bone marrow disorder and intravascular hemolysis into following subgroups.

1. Classic PNH: characterised by overt episodes of intravascular haemolysis.
2. Hypoplastic PNH: in the setting of another marrow disorder, with no overt haemolysis (eg. PNH/aplastic anaemia or PNH /refractory anaemia/myelodysplasia)
3. PNH subclinical (PNH-sc): PNH cells present in the setting of another marrow disorder (eg. PNH-sc/aplastic anaemia). Overt symptoms of PNH are not resent.

Flow cytometry evaluation of erythrocytes and granulocytes provides a rapid, sensitive and specific test for screening and identification of PNH clones. Testing for the absence of GPI-linked antigens on both red cells and granulocytes is recommended for the diagnosis of PNH. It is recommended that two GPI-linked antigens are assessed to confirm PNH as rare hereditary disorders may lack a specific cell surface antigen causing negativity with one antibody and the sensitivity of the test is increased by assessing dual negativity for two GPI linked-antigens.

Large number of GPI-linked antigens can be studied using specific monoclonal antibodies for diagnosis of PNH. FLAER is another reagent being increasingly used for flow cytometric detection of PNH. It is a fluorescently labeled inactive variant of the bacterial protein aerolysin that binds specifically to GPI anchor on cell surface, however does not bind to PNH cells. Although its use is restricted to leukocytes, it is particularly useful for reliable quantification of small PNH clones, as its absence confirms GPI-deficiency.

GPI deficiency on the cells can be partial (type II cells) or complete (type III cells). Cells with normal levels of GPI are referred to as type I cells. PNH type II and type III cells together are considered representative of PNH clones.

Analysis of RBC

Expression of GPI-AP has been extensively studied in RBC as haemolysis is an important clinical manifestation of the PNH disease.

- For routine screening of PNH, it is recommended that simultaneous evaluation of two GPI-linked antigens (CD55 and CD59) using directly conjugated monoclonal antibodies should be performed.
- For establishing the diagnosis, the cells must be GPI-AP deficient.
- Analysis of red cells in an untransfused PNH patient provides the clearest definition of type III (complete deficiency), type II (partial deficiency), and type I (normal expression) populations.

Analysis of granulocytes and monocytes

- Analysis of GPI-AP expression on granulocytes provides accurate clone size of PNH as they are unaffected by red cell transfusion. However standardization of granulocytes staining is technically more difficult.
- For granulocyte analysis, FLAER with CD24/CD16/CD66b/CD55 is recommended.
- For monocytes, FLAER with CD14/CD55 is recommended.
- At least two GPI-linked markers are recommended for granulocyte and monocytes analysis.
- The FSC/SSC gating on granulocytes is often not sufficient in analyzing samples with low granulocyte counts and in small clone sizes. For this purpose, non-GPI linked lineage markers such as CD15, CD33 and CD45 antibodies are useful.
- Analysis of granulocytes should be ideally carried out within the first few hours of collection. As the sample ages, the autofluorescence and the nonspecific binding of the antibodies to the granulocytes increases.

Recommendations for selection of cell populations for analysis:

- Granulocyte analysis is recommended in all cases.
- Monocytes analysis provides confirmatory information.
- RBC analysis should be performed in at least those cases with a PNH clone detected by WBC analysis or in all cases.
- Routine analysis of RBC alone is not recommended.

Reagents

- CD55 PE (clone IA10, BD Biosciences, 555694)
- CD59 FITC (clone p282, BD Biosciences, 555763)
- CD24 PE (clone ML5, BD Biosciences, 555428)
- CD33 PE Cy5.5(clone WM 53, BD Biosciences, 551377)
- FLAER (Protox Biotech, Victoria, BC, Canada, FL2): dissolved in 1mL of Phosphate Buffered Saline (PBS) and "stock" aliquots of 50 µl stored at -200C. "Working" aliquots dilute 1:5 with PBS and stored at 4-8°C and used within 1 week.

- BD FACS lyse solution, (BD Biosciences 349202) 10x. Dilute 1:10 with D/W for working solution.

Reagents can be stored at 4-8°C except for FLAER as described above.

Specimens

2 ml of peripheral blood (A bone marrow sample is not recommended as expression of GPI-linked proteins varies during maturation.) should be collected in EDTA Vaccutainer® tubes. Fresh blood sample is preferred for granulocyte analysis but can be processed up to 48 hrs after collection if stored at 4^0C. For RBC analysis sample can be processed up to 7 days after collection if stored at 4^0C.

Granulocyte Staining Method:

1. Following antibodies are used for immunophenotyping:

No	Tube	PE	FITC	PEcy5
1	**Unstained + CD33**	-	-	CD33
2	**CD24 + FLAER + CD33 (Granulocytes)**	CD24	FLAER	CD33
3	**CD14 + FLAER + CD33 (Monocytes)**	CD14	FLAER	CD33

2. Add 50 µl Blood to the bottom of the tube, taking care not to leaving blood on the sides of the tube.
3. Add 10 µl of each antibody and 10 µl FLAER "Working solution" as mentioned above. Mix well and incubate in dark for 20 min.
4. Add 1 ml BD FACS lyse working solution and incubate for 10 min at room temperature (RT).
5. Centrifuge at 250 RCF for 5 min at RT. Decant the supernatant completely. (Blot the tubes completely dry).
6. Add 4 ml PBS and vortex. Centrifuge at 250 RCF for 5 min at RT. Decant the supernatant.
7. Finally add 200 µl of PBS to each tube. Anaylze and acquire data on 50,000 gated Granulocytes.

RBC Staining Method:

1. Wash 50 µl of blood with 2 ml PBS and dilute (1:150) with PBS and mix well.
2. Take 25µl of diluted sample. Add 10 µl of the CD55 and CD59 antibodies to the diluted sample.
3. Incubate in dark at RT for 15 min, centrifuge and wash twice with PBS and re-suspend in 250 µl of PBS for analysis.

Assay stability

- Neutrophils and monocytes are suitable for testing when stored at 4°C.
- Immunophenotyping: stable up to 24 hrs.
- FLAER assay: stable up to 48 hrs

Note: The antibody combination should be selected and titrated by the laboratory. For these antibody combinations, each laboratory should determine their sensitivity by performing the test on at least 20 normal samples. For routine analysis the recommended sensitivity is >1% and for high sensitivity analysis it is >0.01%.

Set Up and Analysis

Data acquisition is performed immediately after processing the samples using a flow cytometer. Threshold is adjusted in forward scatter (FSC) vs. side scatter (SSC) dot plots to exclude debris. Granulocyte and monocytes populations are identified using populations are identified in the SSC/CD33 dot plots. For each tube, at least 5,000-10,000 gated events are acquired and recorded for routine clinical analysis whereas for high sensitivity analysis, the acquisition of 250,000 events is recommended.

The percentage of Type II and Type III cells should be reported both at diagnosis and follow up on each cell population studied. Classic PNH will usually have >10% of cells from at least 2 lineages in the abnormal clone, unless patient has been recently transfused.

Interpretation

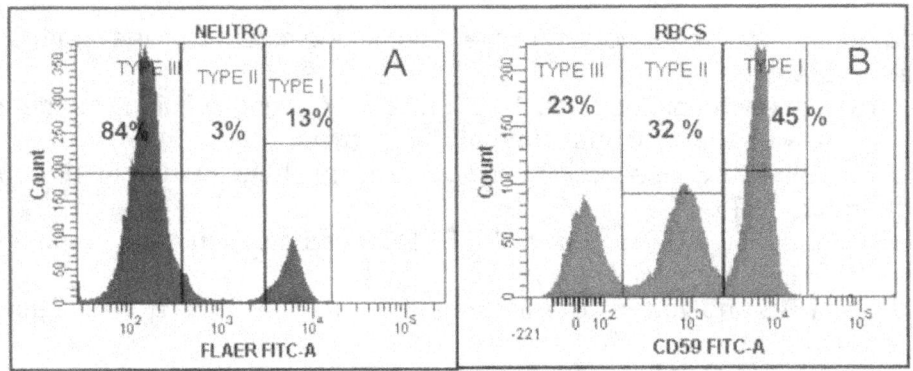

Figure 1. The flow cytometric data shown these histograms supports the diagnosis of PNH with cells of type I (normal expression), type II (partially deficient) cells or type III (completely deficient) present.

Analysis Tips and Hints

- Collect blood sample in sterile Vaccutainer® tubes to avoid neutrophil activation.
- Process the sample as soon as possible after collection.
- There may be a significant technical problem of agglutination while staining RBC with 2 or more monoclonal antibodies. This can be minimized by adequately diluting the cells and adding a wash step with phosphate buffered saline (PBS) before staining.
- CD59 works better than CD55 for RBC analysis. For CD59 clone MEM43 tagged with PE is recommended as FITC labeled clone may cause agglutination of RBC (however not seen with this protocol)

- Addition of CD235a for gating of RBC is recommended for accurate clone size estimation.
- It is important to eliminate dead cells during analysis of granulocytes as they may give false positive (PNH clone +) results.
- The use of FLAER is particularly useful as it provides the best discrimination between PNH cells and cells with normal GPI expression.
- CD59 is not recommended for granulocyte and monocytes analysis as it can give nonspecific positive fluorescence on GPI deficient granulocytes.
- In CD16 analysis, one should keep in mind that normal eosinophils do not expresses CD16 and it can cause confusion and potential misdiagnosis.
- Do not expose FLAER reagent to light as it is photosensitive and the fluorescence intensity of the FLAER will decrease after 6-8 months of reconstitution.
- Recently, a liquid preparation of FLAER has been prepared that shares the binding characteristics of the lyophilized form, but has stability and storage requirements comparable to those of typical monoclonal antibodies. Nonetheless, it is still important to protect this reagent from light and from prolonged exposure to temperatures above 2–8.

References

Borowitz MJ, Craig FE, Digiuseppe JA, Illingworth AJ, Rosse W, Sutherland DR, Wittwer CT, Richards SJ; Clinical Cytometry Society. Guidelines for the diagnosis and monitoring of paroxysmal nocturnal hemoglobinuria and related disorders by flow cytometry. Cytometry B Clin Cytom 2010 78:211-230.

Madkaikar M, Gupta M, Jijina F, Ghosh K. Paroxysmal nocturnal haemoglobinuria: diagnostic tests, advantages & limitations. Eur J Haematol 2009 83:503-511.

Richards SJ, Hill A, Hillmen P. Recent advances in the diagnosis, monitoring, and management of patients with paroxysmal nocturnal hemoglobinuria. Cytometry Part B Clin Cytom 2007 72:291-298.

12. Perforin Assay

Manisha Madkaikar
madkaikarm@icmr.org.in

Introduction
Perforin is a cytoplasmic protein responsible for the translocation of granzyme B from cytotoxic cells into target cells. Granzyme B then migrates to the target cell nucleus to participate in triggering apoptosis. The role of Perforin in immune regulation is still not well defined but, without Perforin, cytotoxic T cells and NK cells show reduced or no cytolytic effect on target cells.

Whole blood is first stained with specific monoclonal antibodies (CD3, CD4, CD8 and CD56) to identify and separate the population of interest. The cells are then fixed, permeablized and stained with a fluorescent labeled monoclonal antibody against Perforin.

About 15-50% of FHLH (Familial Hemophagocytic lymphohistiocytosis) patients show Perforin deficiency. Hence perforin protein expression can be used to identify such patients. However the diagnosis needs to be confirmed with genetic analysis.

Reagents
- CD8PE-Cy7 (clone RPA-T, BD Biosciences,557746)
- CD 56 PE (clone MY31, BD Biosciences, 347747)
- CD 3 PE-Cy5(clone HIT3a, BD Biosciences, 555341)
- CD 4 APC (clone RPA-T4, BD Biosciences, 555349)
- Perforin Kit (BD Biosciences, 556577)
- Cytofix/Permfix Kit (BD Biosciences, 554714) Ready to use.
- Cytofix (BD Biosciences, 51-2090KZ, 554722)
- Perm wash buffer (BD Biosciences, 51-2091KZ, 554723) 10X
- FACS Lyse solution (BD Biosciences, 349202) 10X
- Working FACS Lyse solution: Dilute stock 1:10 with D/W
- Working Perm wash buffer: Dilute stock 1:10 with D/W

All solutions should be stored at 4-8°C.

Specimen
2 ml of peripheral blood should be collected in EDTA Vaccutainer® tubes. A fresh blood sample is preferred, but samples up to 8 hr old can be processed.

Method
1. Add 100 µl of patient blood to the bottom of the tube, taking care not to leave any blood on the sides of the tube.
2. Always run a normal control with a patient sample
3. Add 5 µl CD56-PE, 5 µl CD3-PerCP, 2.5 µl CD8-PECy7 and 2.5 µl CD4-APC to the tube.

4. Vortex and incubate for 15 min in the dark.
5. Add 1 ml of the working FACs lysing solution.
6. Vortex and incubate for 10 min in the dark.
7. Centrifuge at 250 RCF for 5 min at RT. Decant the supernatant completely. (Blot the tubes dry completely)
8. Vortex the tube and add 250 µl of Cytofix/Cytoperm.
9. Incubate for 30 min in the dark.
10. Add 1 ml Wash buffer, vortex and centrifuge at 250 RCF for 5 min at RT. Decant the supernatant.
11. Add 5 µl Perforin-FITC antibody, vortex and incubate in dark for 1 hr.
12. Add 1 ml Wash buffer, vortex and centrifuge at 250 RCF for 5 min at RT. Decant the supernatant.
13. Analyze and acquire data on 5000 NK cells.

Set Up and Analysis

Data acquisition is performed immediately after processing the samples using a flow cytometer and the FACS DIVA software from BD Biosciences. Threshold is adjusted in forward scatter (FSC) vs. side scatter (SSC) dot plots to exclude debris. From each tube at least 5×10^3 (5,000) gated NK cells are analyzed.

Typical results are shown in Figure 1.

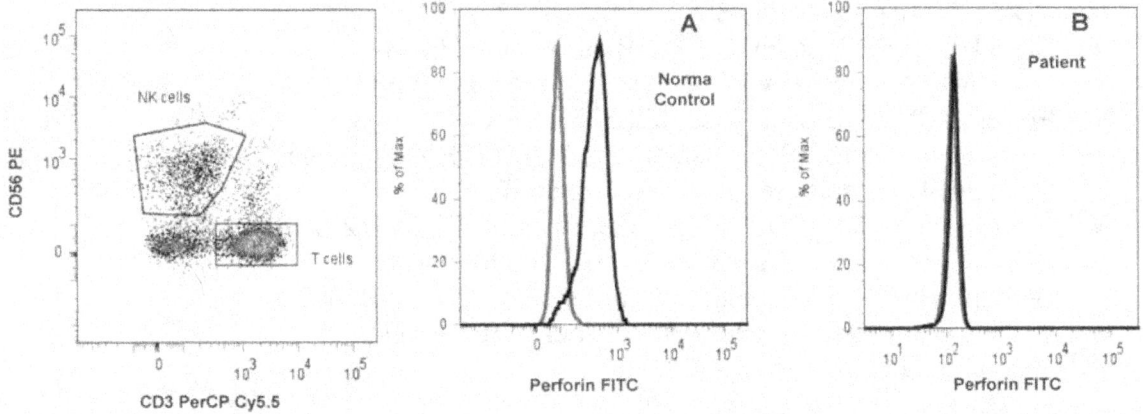

Figure 1. Expression of Perforin on NK cells. A. Normal CD4 cells do not show any Perforin expression (left peak) while a positive control (right peak) does. B. Patient with HLH due to Perforin deficiency shows an absence of Perforin expression (negative control and patient peaks overlap) compared normal control.

Interpretation

- Perforin expression on cytotoxic T lymphocytes and NK cells is seen against CD4 positive cells which act as negative control for diagnosis.

- The results cannot be interpreted if the acquired NK cells count is less than 1000 cells.

Normal range
- Expression on NK cells = 86 ± 5%.
- Expression on Tc cells = 10 ± 5%.

Comments
- RBC lysing step is crucial – Incubate for exactly 10 min.
- Addition of cytofix/cytoperm reagent requires vortexing   of tube for proper mixing.

References

Katano H, Cohen JI. Perforin and lymphohistiocytic proliferative disorders. Br J Haematol. 2005 128:739-50.

Stepp SE, Dufourcq-Lagelouse R, Le Deist F, Bhawan S, Certain S, Mathew PA, Henter JI, Bennett M, Fischer A, de Saint Basile G, Kumar V. Perforin gene defects in familial hemophagocytic lymphohistiocytosis. Science 1999 286:1957–1959.

13. Bruton's Tyrosine Kinase Analysis

Manisha Madkaikar
madkaikarm@icmr.org.in

Introduction

X-linked agammaglobulinemia (XLA) is a primary immunodeficiency disease caused by mutations of Bruton tyrosine kinase (Btk) gene resulting in absent or reduced expression of protein Btk in B cells. It is characterized by a paucity of circulating B cells and a significant reduction in the serum immunoglobulin concentrations that predispose the affected patients to frequent and severe bacterial infections. Apart from B lymphocytes, Btk is highly expressed by monocytes while granulocytes and T-lymphocytes show weak expression of Btk. Using an anti-Btk monoclonal antibody, intracytoplasmic Btk protein expression on monocytes can be studied by flow cytometric analysis. Thus the demonstration of absence or reduced Btk expression, along with classical clinical presentation, is diagnostic of XLA. Demonstration of BTK mosaicism in monocytes may also help in the detection of obligate XLA carriers. However the diagnosis needs to be confirmed with genetic analysis. Patients diagnosed with this condition are placed on lifelong immunoglobulin replacement.

Reagents

- Formaldehyde (Breon) – 35%
- Triton X (Qiagen)
- PBS Ca+ and Mg+ free (Gibco, 10010)
- Anti-Btk Alexa 647 (clone53/BTK, BD, 558528)
- Anti-CD14 PE (Clone MΦP9, BD, 347497)
- Anti-CD3PerCP (HT3a, BD Pharmingen, 555341)
- Bovine Albumin (Sigma, A-3059)
- 10% Formaldehyde - Add 285 µl of 35% formaldehyde to 715 µl PBS. Prepare fresh daily.
- Wash buffer: To 50 ml of PBS, add 0.5 g BSA (making a 1% solution). Invert to mix, let it stand for approximately 10 minutes for the BSA to dissolve, then invert to mix again. Wash buffer is stable in the fridge for 1months.
- Wash buffer with Triton X: Add 50 µl of Triton X to 10 ml of wash buffer (PBS+ 1% BSA) (making 0.05% Triton X solution). Invert to mix.
- 1% Triton X solution: To 10 ml of PBS, add 100 µl Triton X (making a 0.1% Triton X solution). Invert to mix. Pre-warm at 37°C for 30 min.

Specimen

2 ml of peripheral blood should be collected in EDTA Vaccutainer® tubes. A fresh blood sample is preferred, but samples up to 8 hr old can be processed.

Method

1. Label each tube with the control or patient's initials.
2. Antibody panel to be used is shown below.

Normal Control	Patient	Tube Name
CD14 PE/ CD3PerCP/ Anti-BTK	CD14 PE/ CD3PerCP/ Anti-BTK	Test
CD14 PE/ CD3PerCP	CD14 PE/ CD3PerCP	Unstained

3. Add 100 µl of whole blood to the bottom of the tube, taking care not to leave blood on the sides of the tube. Then add 65 µl of 10% formaldehyde. Vortex immediately and incubate at RT for 10 minutes.
4. Add 1 ml of a 0.1%, pre-warmed Triton X solution (37^0C) to all the tubes, vortex to mix well, and incubate at 37^0C for 30 minutes.
5. Add 2 ml of wash buffer (cold) to each tube and spin at 690 RCF for 10 min at 4^0C.
6. Decant properly, add 2 ml of wash buffer with Triton X (cold) to each tube and spin at 690 RCF for 10 min at 4^0C
7. Decant properly and resuspend in 50 µl of wash buffer with Triton X.
8. To the "Test" tubes, add 5 µl of anti-BTK + CD3 + CD14.
9. To the "Unstained" tubes add 5 µl of CD3 + CD14 without anti-BTK.
10. Gently mix all tubes and incubate for 1 hr at RT.
11. Wash once with PBS.
12. Analyze and acquire 5,000 monocyte events. Samples can be analyzed immediately or up to 96 hours later.

Set Up and Analysis

Data acquisition is performed immediately after processing the samples using flow cytometer and the FACS DIVA software. Threshold is adjusted in forward scatter (FSC) vs. side scatter (SSC) dot plots to exclude debris. For each tube at least 5×10^3 (5,000) gated monocytes are analyzed. A normal control is processed along with each sample. Typical results are shown below in Figure 1.

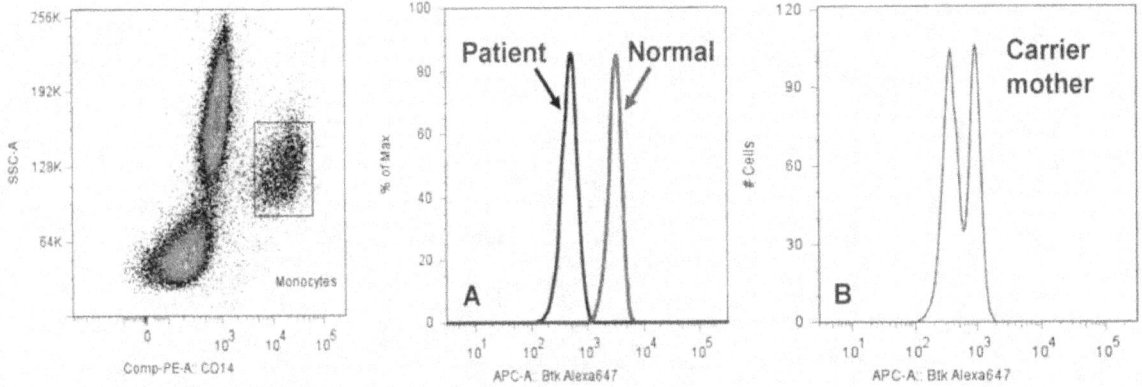

Figure 1: Expression of Btk in gated monocytes: A. Shows absent expression of Btk on monocytes (left) of patient with XLA as against normal (right). B. Shows bimodal pattern of Btk expression on monocytes suggesting carrier status.

Interpretation
- Btk expression in monocytes is seen against CD3 positive cells which act as negative control for diagnosis.
- Normal range = 90 ± 5%.
- Carrier = a distinct positive and negative peak is seen.

Comments
- Collect blood sample in sterile Vaccutainer® tubes to avoid activation of monocytes.
- Process sample as soon as possible after collection.
- Fixation step with formaldehyde is crucial – Incubate for exact 10 min. Prepare freshly.
- Accurate concentration of Triton X is very crucial for optimum staining.
- All clones of CD14 do not withstand the permeabilization procedure, hence selection of proper clone is essential
- All B lymphocytes in Carrier patients are positive for Btk as the deficient lymphocytes do not survive; hence it is essential to study the Btk expression on monocytes for Carrier Detection.

References
Futatani T, Miyawaki T, Tsukada S, Hashimoto S, Kunikata T, Arai S, Kurimoto M, Niida Y, Matsuoka H, Sakiyama Y, Iwata T, Tsuchiya S, Tatsuzawa O, Yoshizaki K, Kishimoto T. Deficient expression of Bruton's tyrosine kinase in monocytes from X-linked agammaglobulinemia as evaluated by a flow cytometric analysis and its clinical application to carrier detection. Blood 1998 91:595-602.

14. Side Population (SP) labeling with DyeCycle Violet (DCV)

William Telford
telfordw@mail.nih.gov

Introduction
Several excellent review articles cover identification of side population or SP cells by staining with Hoechst 33342 (Montanaro et al., 2004; Goodell, 2005; Lin and Goodell, 2006; Petriz, 2007). The original paper by Margaret Goodell is also still relevant (Goodell et al., 1996). Telford et al introduced DyeCycle Violet as a violet laser excited dye for the detection of SP cells thus obviating the need for use of Hoechst 33342 and the UV laser excitation for detection of SP cells (Telford et al. 2007 and Telford 2009). Like Hoechst 33342, the use of DyeCycle Violet (DCV) for SP labeling is very sensitive to labeling conditions and the following instructions should be closely followed.

Reagents
The following reagents should be prepared in advance:
- Hoechst 33342 powder (Invitrogen, Sigma Chemical Co. and other sources). Hoechst 33342 should be prepared as a 1 mg/ml stock solution in distilled water (precipitation can occur over time in buffers). Stock can be stored at 4°C for up to one month. Powdered Hoechst 33342 should be stored over desiccant for up to one year.
- Efflux inhibitors: An ABCG2 pump inhibitor should be added to a control sample to allow accurate gating of the SP population. Effective inhibitors include verapamil (Sigma Chemical Co.) and fumitremorgin C. For verapamil, prepare a 10 mM stock solution in DMSO, for final dilution to 50 µM. For fumitremorgin C, prepare a stock at 2 mM in distilled water, for final dilution to 10 µM. Both inhibitors can be stored in frozen aliquots.
- Wash buffer (HBSS+): HBSS (without phenol red), supplemented with 2% fetal bovine serum in 2 mM HEPES.
- SP buffer (DMEM+): DMEM (no phenol red) with 2% fetal bovine serum and 2 mM HEPES.
- Propidium iodide: This is used as a viability label to exclude dead cells from the analysis. Prepare a 1 mg/ml stock in PBS, for final dilution to 2 µg/ml.
- DyeCycle Violet (DCV) (Invitrogen Life Technologies): DCV can be used in place of Hoechst 33342 when a violet laser is available. DCV is supplied as a 5 mM solution, for final dilution to 5-10 µM.
- A 37°C water bath, centrifuge and source of ice for 4°C storage.

Method
DCV SP Labeling
1. Prepare the cells to be labeled in SP (DMEM+) buffer, at concentrations up to 5 x 10^6 cells/ml. Cells can be labeled at lower concentrations if necessary. Use HBSS+ buffer for preparations steps, but resuspend in DMEM+ at the final step.

2. Divide the cell sample into two tubes, one with no inhibitor, and the other as an inhibitor control. Warm the samples to 37°C in a water bath prior to labeling
3. Add inhibitor (verapamil or fumitremorgin C) to the control sample. At the concentrations given above (10 mM and 2 mM respectively), this is a 1:200 dilution for both inhibitors (5 μl per 1 ml). Incubate at 37°C for 15 minutes.
4. After preincubation with inhibitor, add 2 μl of DCV (5 mM stock solution) per 1 ml of cells. Incubate for 90 minutes at 37°C with gentle mixing at 30 minute intervals.
5. Centrifuge the sample tubes at 400 X RCF for 7 minutes at 4°C, decant and resuspend in cold DMEM+ in the same volume as the labeling. The tubes can be stored for up to four hours on ice, but should be analyzed as quickly as possible. Add propidium iodide at 2 μg/ml final concentration (2 μl of a 1 mg/ml stock solution per 1 ml cell suspension) a few minutes before analysis.

DCV SP Labeling and Simultaneous Immunophenotyping

As with Hoechst 33342, DCV SP labeling is compatible with simultaneous immunolabeling for stem cell surface markers. Unlike Hoechst 33342, however, DCV is somewhat excited by the 488 nm laser, causing some emission in the FITC and PE range in DCV labeled cells. These fluorochromes should therefore not be used with DCV labeled cells. However, PE-Cy5, PE-Cy5.5, PE-Cy7, APC, APC-Cy5.5 and APC-Cy7 are all spectrally compatible with DCV labeling. Stem cell antibodies are now available as direct conjugates for all of these fluorochromes. Simultaneous immunophenotyping should be carried out after DCV labeling, once the cells are on ice.

1. Following incubation with DCV and washing, resuspend the cells in 200 to 500 μl of HBSS+ and place on ice.
2. Add pre-titered antibody, and incubate at 4°C for 15-30 minutes. Add 3 mls of cold HBSS+ buffer and centrifuge at 400 x g for 5-7 minutes at 4°C.
3. Repeat the above steps for additional antibodies if necessary. For human cells, antibody labeling can usually be done in one step. For mouse cell labeling in several steps separated by centrifuge washes may be necessary.
4. After the cells are labeled, return the suspension to its original volume with DMEM+ and store at 4°C until analysis. As above, add propidium iodide 5 minutes prior to analysis.

Flow Cytometric Analysis

A flow cytometer with a violet or ultraviolet laser is required for DCV SP analysis. The BD Biosciences FACSCanto series (I and II), LSR II, LSR Fortessa, FACSAria series (I, II and III), the Beckman-Coulter Gallios, the Stratedigm S-series and the Partec CyFlow instruments are frequently equipped with violet and sometimes ultraviolet lasers. Before preparing cells, make certain your instrument has a violet laser, and that it is equipped with the necessary detectors (two required) and filters.

1. Set up the instrument for analysis. For DCV, either a UV or violet laser can be used, with two aligned detectors (often set up for Pacific Blue and Pacific Orange on many commercial instruments). Examples are shown in Figure 1.

2. Laser alignment should be checked prior to cell analysis as a poorly aligned laser can have a significant effect on SP resolution. UV or violet laser alignment can be checked with several microbead standards, including Spherotech Ultra Rainbow microspheres (Spherotech, Libertyville, IL). These microspheres are well-excited by both UV and violet lasers, and can be used for daily quality control of your instrument. InSpeck Blue microsphere arrays (Invitrogen Life Technologies, Carlsbad, CA) are a mixture of seven bright-to-dim microspheres; they are even more useful for identifying minor degradation in instrument alignment. InSpeck Blue spheres will be better excited by UV than violet lasers.

3. The violet-aligned detectors should have blue and red narrow bandpass filters inserted. Any Pacific Blue or DAPI bandpass filter (i.e. 450/40 nm, 440/10 nm, etc.) will work for the Hoechst or DCV blue signal, and any APC or Cy5 filter (675/20 nm, 660/20 nm) will work for the red. Depending on the instrument design, a short pass or long pass dichroic mirror ranging from 560 to 610 nm will work to split the signals. A 580 nm long pass mirror is typically used on a BD Biosciences LSR II, FACSCanto II or FACSAria II. A ~600 nm shortpass filter is used on a Beckman-Coulter Gallios (Figure 1).

4. Set both DCV blue and red detectors for linear acquisition (they may normally be set to log scaling).

5. Run the cells on the flow cytometer, and display them in a forward versus side scatter two-parameter dot plot. Bone marrow can have a multi-population forward versus side scatter profile, so ensure all populations of interest are on scale. Gate on the scatter populations of interest. Mouse bone marrow is shown in Figure 2.

6. Then create a side scatter versus PI fluorescence dot plot, and draw a gate for the gate on the PI-negative cells. The viable cell background fluorescence in the PI detector will be somewhat higher with DCV than that normally observed for Hoechst 33342, since DCV is somewhat excited at 488 nm. However, it should be possible to distinguish PI-negative viable cells prom PI-PI-positive apoptotic and necrotic.

7. Finally, display a DCV red (X-axis) versus DCV blue (Y-axis) dot plot, gated for scatter and PI viability. Adjust the voltages on the blue and red detectors to place the dominant G_1 population roughly in the center of the dot plot. For DCV, the SP population should arch up along the Y-axis, and eventually curve back down toward the minimum points of both axes. This separation will be much more pronounced than that normally observed for Hoechst 33342 (Figure 2 shows mouse bone marrow, human bone marrow and human cord blood).

8. Check all phenotypic markers and set their detector voltages appropriately. When setting compensation, ignore the DCV blue and red signals; very little spectral overlap should occur if you are not using fluorescein or PE. When using automated spillover routines, exclude DCV blue and red signals from the calculation, since the software will attempt to compensate DCV blue and red overlap from, giving an altered labeling pattern).

9. In most hematopoietic tissues, the SP population will be rare, usually less than 0.1% of the viable nucleated cells. Collect at least 500,000 events per sample if possible.

10. Use the inhibitor control to set the upper limit for the SP. With DCV, this cutoff may be lower on the SP arch than with Hoechst 33342. In many cases, verapamil and fumitremorgin C may not completely inhibit dye efflux in the SP population.

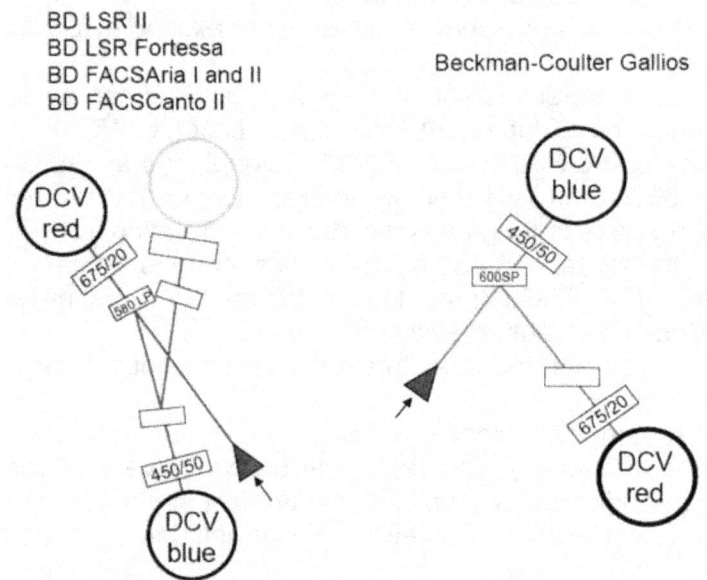

Figure 1. Filter/detector combination for BD (left) and Beckman-Coulter (right) flow cytometers.

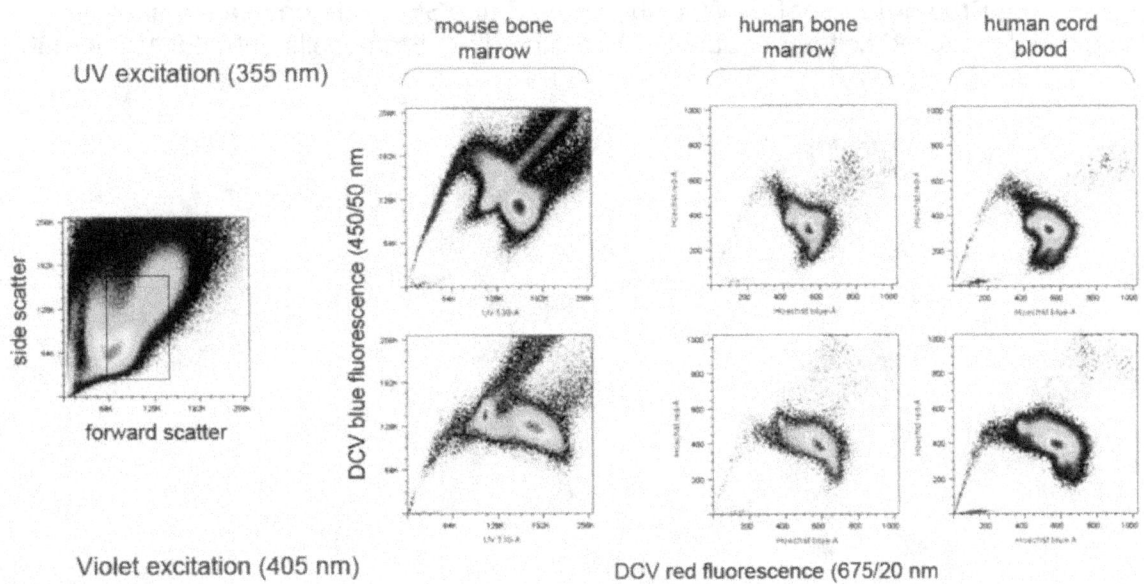

Figure 2. Left, scatter settings for mouse bone marrow. Right, DCV SP analysis for mouse bone marrow, human bone marrow and human cord blood, analyzed with an ultraviolet laser (upper row) or violet laser (lower row).

References

Goodell, MA. 2005. Stem cell identification and sorting using the Hoechst 33342 side population (SP). Current Protocols in Cytometry 2005 Unit 9.18. pp9.18.1-9.18.11.

Goodell MA, Brose K, Paradis G, Connor AS, Mulligan, RC. Isolation and functional properties of murime hematopoetic stem cells that are replicating in vivo. J Exp Med 1996 183:1797-1806.

Montanaro F, Liadaki K, Schienda J, Flint A, Gussoni E, Kunkel LM. Demystifying SP cell purification: viability, yield, and phenotype are defined by isolation parameters. Exp Cell Res 2004 298:144-154.

Lin KK, Goodell MA. Purification of hematopoietic stem cells using the side population. Methods in Enzymology 2006 420:255-264.

Petriz J. Flow cytometry of the side population (SP). Current Protocols in Cytometry 2007 Unit 9.23, pp9.23.1-9.23.14.

Telford WG. Stem cell side population analysis and sorting using DyeCycle Violet. Current Protocols in Cytometry 2009 Unit 9.30, pp9.30.1-9.30.9.

Telford WG, Bradford J, Godfrey W, Robey RW, Bates SE. Side population analysis using a violet-excited cell permeable DNA binding dye. Stem Cells 2007 25:1029-1036.

15. Flow Cytometric Enumeration of CD34+ Hematopoietic Progenitor Cells According to ISHAGE Protocol

Michael Keeney
Mike.Keeney@LHSC.ON.CA
Robert Sutherland
rob.sutherland@utoronto.ca

Introduction
The ISHAGE protocol uses CD34 and CD45 staining together with a sequential Boolean gating strategy to delineate and enumerate hematopoietic CD34+ cells (Sutherland et al., 1994, 1996). By incorporating a known number of fluorescent counting beads and assessing the ratio between the number of beads and CD34+ cells, an absolute CD34+ cell count can be generated using a flow cytometer alone (single-platform methodology), (Keeney et al.,1998; Brocklebank and Sparrow, 2001; Keeney and Sutherland, 2007). Simple ammonium-chloride-based lyse-no-wash sample processing is used. Finally, the inclusion of the viability dye 7-aminoactinomycin D (7-AAD) allows the exclusion of dead cells and the enumeration of viable (v)CD34+ cells (Keeney et al.,1998). The ISHAGE protocol can be used on a variety of cytometers and with a variety of validated antibody conjugates and fluorescent counting beads (Sutherland et al., 2009).

Equipment
- Electronic pipettor capable of reverse pipetting.
- 12 × 75 mm polystyrene tubes.
- Pipette tips.
- Flow cytometer.
- Vortex mixer.

Supplies and Reagents
- Phosphate-buffered saline (PBS) with 1% (w/v) human serum albumin.
- Ammonium chloride-based red cell lysis buffer: (e.g., PharMLyse, BD Biosciences, Stemlyse, Beckman Coulter StemKit).
- Anti-CD34-PE: (e.g., 581 Beckman Coulter, Franklin Lakes, NJ or HPCA-2 clone, BD Biosciences, San Jose, CA) or validated alternative.
- Anti-CD45-FITC or anti-CD45-PerCP: (e.g., clone J33 (Beckman Coulter or HLE-1, BD Biosciences) or validated alternative.
- 7-AAD: (e.g. ViaProbe, BD Biosciences).
- Fluorescent beads for single-platform assay (e.g., Flow-Count Fluorospheres, Beckman Coulter).

Method
1. Create acquisition/analysis template on flow cytometer.
   Note: Beckman Coulter has developed a reagent and software package that contains all the necessary components for sample preparation, instrument setup,

data acquisition, analysis and color compensation. This package was developed for the Coulter XL analyzer (StemOne) and more recently for the FC500 instrument (StemCXP) (see Figure 1). BD Biosciences CellQuest software (version 4.0 and above) contains expression editors that can automate the analysis regardless of the counting beads used (see Figures 2 through 5). The following setup shows how to create the protocol for analysis on BD instruments. More information on templates for Beckman Coulter instruments is found in the "Notes" section.

a) Create the following bivariate dot plots as shown in Figure 1:
   1. Plot 1: CD45-FITC (FL1) vs. side scatter (SS).
   2. Plot 2: CD34-PE (FL2) vs. SS.
   3. Plot 3: CD45-FITC (FL1) vs. SS.
   4. Plot 4: Forward scatter (FS) vs. SS.
   5. Plot 5: CD45-FITC (FL1) vs. CD34-PE (FL2).
   6. Plot 6: FS vs. SS.
   7. Plot 7: Time vs. FS (for BD FACS instruments if Flow-Count beads are used). (If Trucount tubes are used, this plot is not required. See below for specific modifications required for the Trucount-tube based assay shown in Figure 4.)
   8. Plot 8: 7-AAD vs. SS.
   9. Plot 9 (optional): 7-AAD vs. SS (for R1 + R2 + R3). This plot is useful for analysis of post-thawed samples and some fresh samples containing a significant number of dead/apoptotic cells. Figure 3 shows the value of such an approach using the same list mode file as used in Figure 2 without viability discrimination (only plots 4, 6, 8, and 9 are shown).

b) Create gating regions and logical gates: In the following description, specific terminology for gating regions (e.g., R1) and logical gates (e.g., G1) are for BD Biosciences FACS instruments.
   - Logical gate setup for BD template using Flow-Count fluorospheres (Figures 2 to 5):
     1. Gate 1 (vCD45): R1 and R8.
     2. Gate 2: R2 and G1.
     3. Gate 3: R3 and G2.
     4. Gate 4: (vCD34): R4 and G3.
     5. Gate 5 (singlet beads): R6 and R7.
     6. Gate 6 (all CD34): R1 and R2 and R3.
     7. Gate 7 (v lymphs): R5 and R8.
     8. Gate 8 (all CD45): R1 and not R6.
     9. Gate 9 (all beads): R6.
   - Plot 1 (leukocyte gate): Display all events. Draw a rectangular region (R1) to include all CD45 events (from dim to bright) and to exclude debris, platelets, and unlysed erythrocytes, which are all CD45–. The counting beads are found in the brightest FL1, FL2, and FL3 channels and also exhibit very high SS. Because the gate statistics are obtained from events in plot 1, it is critical that region R1 includes the highest FL1 and SS channels.

- Plot 2 (total CD34 gate): Display events from the leukocyte gate (gate G1 = R1). Draw an amorphous polygon region (R2) to include all CD34+ events.
- Plot 3 (CD34+ blast gate): Display events that fulfill the criteria of both preceding gates (Regions R1 and R2; G2 = R2 and G1). Draw an amorphous polygon region (R3) to include only those events that form a cluster with low to intermediate SS and dim CD45 expression.
- Plot 4 (lymph-blast gate): Display events that fulfill the criteria of all three preceding gates (Regions R1, R2, and R3, G3 = R3 and G2). Draw an amorphous polygon region (R4) to include only those events that form a cluster with low to intermediate SS and low to intermediate FS. Set logical gate G4 = R4 and G3. The lymph-blast gate serves to exclude platelets and debris that may show weak nonspecific binding of CD34 and CD45. Its optimal positioning and lower boundaries are set as described in plot 6.
- Plot 5: Display ungated data. Draw a quad-stat region to establish the lower limit of CD45 expression by the CD34+ events. On plot 5, draw a small rectangular region (R6) to include the brightest events that fall in the highest FL1 and FL2 channels. Although not readily visible, all of the Flow-Count beads (singlets and aggregates) are contained in this region. A rectangular region is established before list mode data acquisition to exclude CD34–/CD45– debris. This is particularly useful for any sample containing high red cell content, and much debris can be thereby excluded from acquisition.
- Plot 6 (duplicate lymph-blast gate): On plot 1, draw an amorphous polygon region (R5) to include the lymphocytes (bright CD45, low SS) to create a lymphocyte gate. Display the events from this region on plot 6 gated on G7 (logical gate G7 = R5 and R8). Copy region R4 from plot 4 into plot 6 (creating the duplicate lymph-blast gate), and adjust the position of the duplicate so that the smallest lymphocytes from region R5 are just included. The original region R4 on plot 4 will automatically move to the same position as that on plot 6.
- Plot 7 (bead gate): Display events from R6 on plot 7 (time vs FS). If beads are not visible, lower FS threshold until all singlet beads can be acquired. Set region R7 to include only singlet beads. Set logical gate G5 = R6 and R7.
- Plot 8: Draw a rectangular region (R8) to include only living (7-AAD–) cells. Display data from region (R8) on plots 1-4 and 6 (see logical gates above). Plot 5 should remain ungated.
- Plot 9 (optional, see Figure 3): Duplicate rectangular region (R8) and display data from regions R1, R2, and R3 only. This plot shows all CD34+ cells regardless of viability and is useful for the accurate placement of the viability gate R8 in post-thawed samples (see Figure 5) or other samples containing large proportions of nonviable cells. The detection of the viable cells on-scale in the first decade or so of

fluorescence also confirms the appropriate instrument setup and compensation between the PE and 7-AAD channels of the cytometer.

c) Make specific modifications required for Trucount-tube-based protocols (Figures 4 and 5):

- Because of the very small size of Trucount beads, a threshold is set in the FL1 channel rather than on FS. With an appropriately set FL1 threshold (see plot 1, Figures 4 and 5), much of the CD34–/CD45– debris is excluded from listmode file acquisition. However, some small debris is usually stained by CD45- FITC, and to exclude it, a small, rectangular exclusion gate (region R7) is established on plot 6. Although this debris can be removed during analysis, excluding it during acquisition and further optimizing it if necessary during analysis is most advantageous, as was performed for the data file shown in Figures 4 and 5. This modification greatly increases the accuracy of the absolute CD45+ cell count. The latter, when compared to the absolute white cell count derived from a hematology analyzer, represents an important quality assurance characteristic of the methodology, especially if only a single tube is stained.
- Trucount beads are gated on plot 5 within an amorphous region R6. Because the assayed bead content of Trucount tubes is determined on all beads (singlets and aggregates), it is not necessary to further analyze the events therein. Logical gate setup for BD Trucount analysis template (Figure 4):

    1. Gate 1 (vCD45): not R7 and R1 and R8.
    2. Gate 2: R2 and G1.
    3. Gate 3: R3 and G2.
    4. Gate 4 (vCD34): R4 and G3.
    5. Gate 5 (beads): R6.
    6. Gate 6 (all CD34): R1 and R2 and R3.
    7. Gate 7 (v lymphs): R5 and R8
    8. Gate 8 (all CD45): R1 and not R6 and not R7.
    9. Gate 9 (debris): R7.

2. Stain samples.
   a) If necessary, dilute specimen with PBS containing 1% serum albumin to obtain a total nucleated cell count of 10 to 20 × 10^9/L.
   b) To a 12 x 75 mm test tube, add CD45FITC/CD34PE (appropriately titrated) and 7-AAD (final concentration = 1 µg/ml).
   c) Accurately pipet 100 µl of well-mixed specimen into the very bottom of the tube and mix. This step requires the use of positive displacement techniques or electronic pipettors with reverse pipetting capabilities.
   d) Incubate tubes at room temperature for 20 minutes, protected from the light.
   e) Add 2 ml of freshly diluted (1:10) ammonium chloride lysing solution at room temperature and vortex to mix.

Note: When post-thawed samples are stained, this step is not necessary. PBS can be used instead, and after addition of the beads (step h), the sample can be acquired immediately.

    f) Incubate 10 minutes at room temperature in the dark.

    g) Following incubation, keep samples on ice and in the dark until flow analysis.

    h) Immediately before analysis, accurately pipet 100 µl of properly suspended Flow-Count Fluorospheres to the tube using the same pipet as above.

    i) Cap tube and gently mix by inversion.

3. Perform flow acquisition and analysis.

    a) Once fluorescent beads are added, acquisition should take place immediately or after a maximum of 1 hour on ice.

    b) A minimum of 75,000 CD45+ events should be collected with a minimum of 100 CD34+ cells to maintain a coefficient variable of 10%.

Anticipated Results

The lower limit of sensitivity is partially determined by the number of CD34+ events collected. The method is sensitive at least to a level of 5 CD34+ cells per µL. Increasing the total number of CD34+ events collected will allow for accuracy below this value.

Notes

- There are specific issues with Flow-Count beads on Beckman Coulter vs. BD FACS instruments. On Beckman Coulter instruments, the FS threshold is set just below the smallest lymphocytes. On BD FACS series instruments, a similar setting would exclude the Flow-Count beads from acquisition. Thus, the FS has to be reduced to a point below where singlet Flow-Count beads are detected on the FS vs. time plot (plot 7).

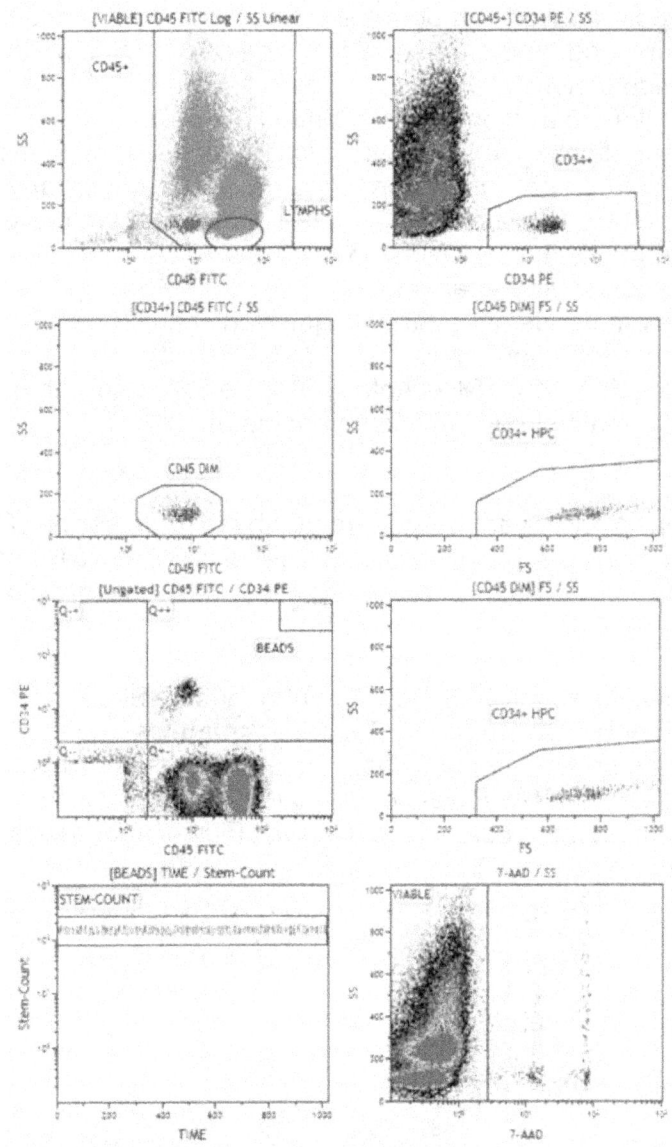

Figure 1. Enumeration of viable CD34+ cells with Stem-Kit™ and Automated Stem-CXP™ software analysis on a Beckman Coulter FC 500. Fresh apheresis sample diluted 1/10 with PBS/1%BSA. Histogram 1 - all CD45+ leukocytes, gated on 7- AAD negative (viable) events from histogram 5. Histograms 2-4 are sequentially Boolean gated from histogram 1. Histogram 6 is gated on lymphocytes from histogram 1 to allow the discriminator to be set on forward scatter. Histogram 7 shows Stem-Count beads (FL3) versus time. Histogram 8 shows singlet beads captured in the top right hand corner and a "live gate" in the bottom left corner, which is used to exclude debris from the listmode file, particularly important in cord blood, PB or bone marrow files. The leukocyte count was 18.3 x 10^9/L, viable CD34+ percent and absolute count were 0.77% and 608/uL respectively (from histogram 4). The sample was 94.9% viable (from histogram 5).

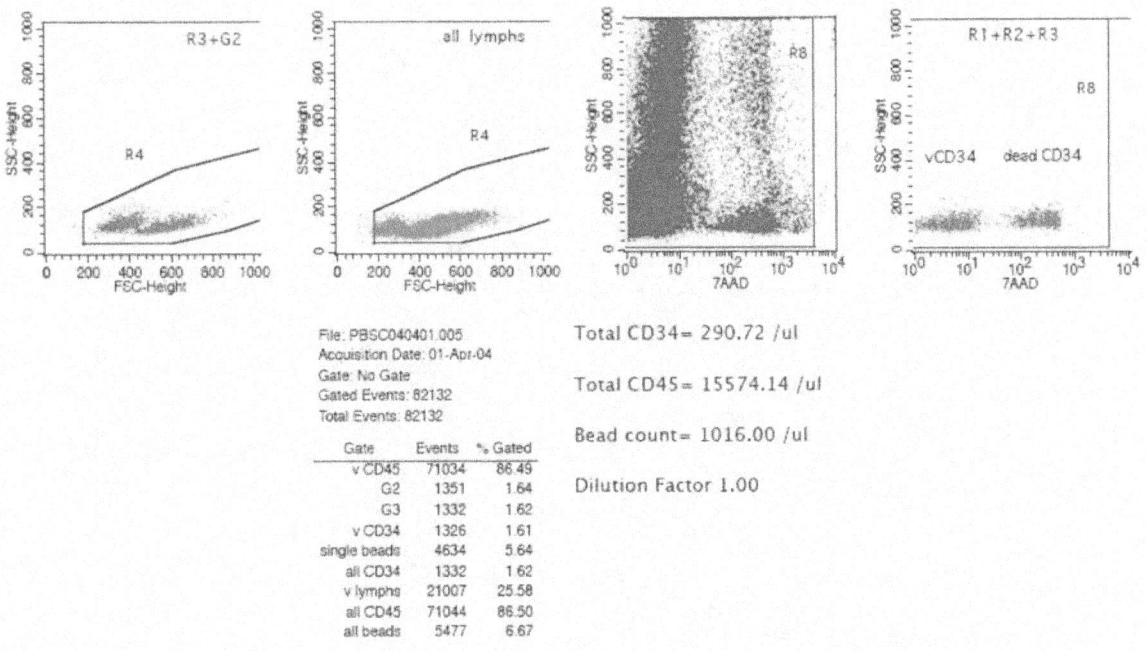

File: PBSC040401.005
Acquisition Date: 01-Apr-04
Gate: No Gate
Gated Events: 82132
Total Events: 82132

Gate	Events	% Gated
v CD45	71034	86.49
G2	1351	1.64
G3	1332	1.62
v CD34	1326	1.61
single beads	4634	5.64
all CD34	1332	1.62
v lymphs	21007	25.58
all CD45	71044	86.50
all beads	5477	6.67

Total CD34 = 290.72 /ul

Total CD45 = 15574.14 /ul

Bead count = 1016.00 /ul

Dilution Factor 1.00

Figure 2. Enumeration of viable CD34+ cells with Stem-Kit on a BD Biosciences FACSCalibur cytometer equipped with CellQuestTM. An apheresis sample that had been stored overnight at room temperature was stained with the Stem-Kit reagent set. Viable CD34+ cells were identified using Boolean gating and regions R1 through R4 (plots 1-4), including only viable (7-AAD-) cells from region R8 (plot 8). Viable lymphocytes from region R5 (plot 1) and R8 are displayed on plot 6 and the duplicate blast-lymphocyte region R4 adjusted to include the smallest viable lymphocytes. Duplicate gating region R4 on plot 4 self-adjusts accordingly. Plot 5 shows the position of a 'live' gate in the bottom left corner, which excludes debris resulting from lyse-no-wash sample processing of PB, CB and BM sample types. The number of CD34+ cells in region R4 is compared with the total number of singlet beads counted in the same listmode file. In the example shown, total beads are gated in Region R6 on plot 5 and displayed on plot 7 (time versus forward scatter). Singlet beads are then delineated and enumerated in gating region R7. Sample analysis was performed using Cellquest ProTM software using semi-automated Expression Editors. For earlier versions of Cellquest, the absolute number of viable CD34+ cells/µl is calculated as follows:

$$\frac{\text{\#CD34+ cells x bead concentration x DF}}{\text{\#singlet beads}}$$

Where #CD34+ cells is determined from logical gate G4 (vCD34 in gate stats = R1+R2+R3+R4+R8), the bead concentration is specified by the manufacturer, DF is the sample dilution factor, and the singlet bead count is determined from plot 7 (singlet beads in gate stats = R6+R7). Plot 9 shows total CD34+ cells (viable and non-viable) from gating regions R1+R2+R3 only and shows viable cells on-scale in about the first decade of fluorescence. This plot is useful when samples with poor viability are to be analyzed as it is easier to set region R8 on this plot versus plot 8. Additionally, it shows that the fluorescence compensation between PMT 2 (CD34PE) and PMT 3 (7-AAD) is optimally set.

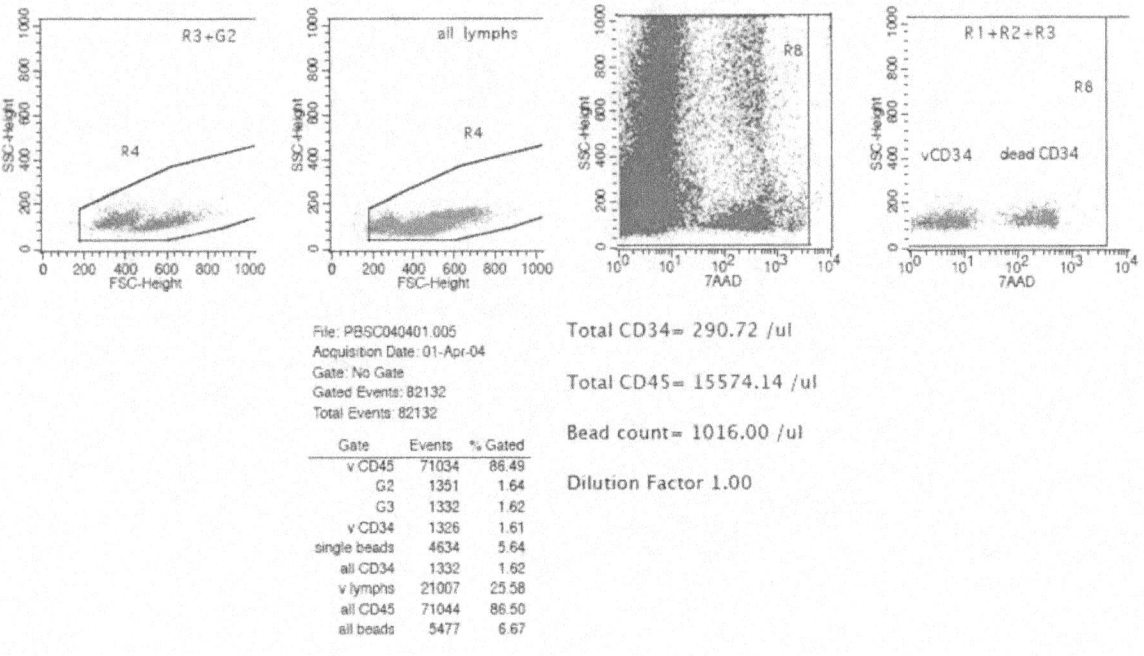

File: PBSC040401.005
Acquisition Date: 01-Apr-04
Gate: No Gate
Gated Events: 82132
Total Events: 82132

Gate	Events	% Gated
v CD45	71034	86.49
G2	1351	1.64
G3	1332	1.62
v CD34	1326	1.61
single beads	4634	5.64
all CD34	1332	1.62
v lymphs	21007	25.58
all CD45	71044	86.50
all beads	5477	6.67

Total CD34= 290.72 /ul

Total CD45= 15574.14 /ul

Bead count= 1016.00 /ul

Dilution Factor 1.00

Figure 3. Importance of viability dye (7-AAD) inclusion in the analysis of non-fresh samples. Analysis of the same sample as Figure 2, except that viability discrimination with 7-AAD has NOT been applied. When gating region R8 is expanded to include both viable and non-viable cells (plots 3 and 4), and region R4 is moved to include both live and dead lymphocytes (plot 2), both dead and live CD34+ cells now cluster within the duplicate lymph-blast region R4 on plot 1. Both the absolute CD34 and CD45 counts are significantly increased versus the values obtained in Figure 2. Note the extra population of both CD34+ cells (plot 1) and lymphocytes (plot 2) with reduced forward angle light scatter; these are the dead cells.

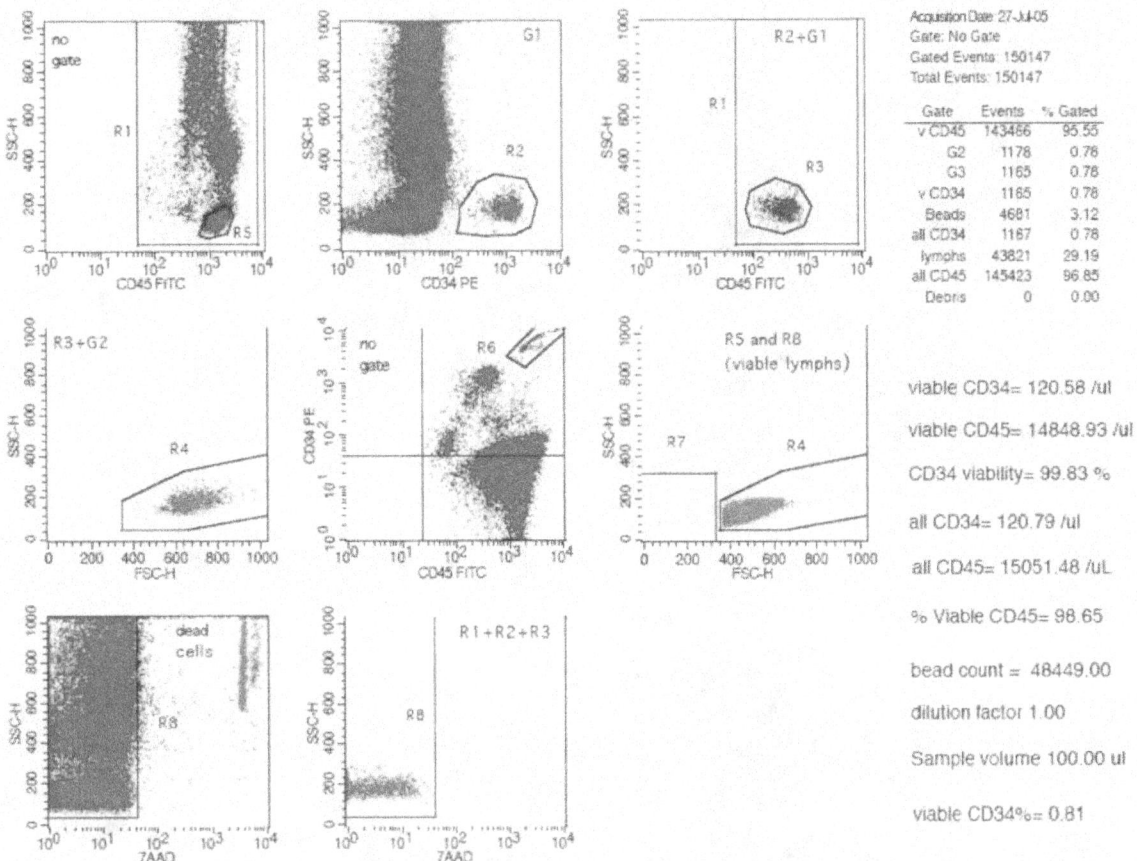

Acquisition Date: 27-Jul-05
Gate: No Gate
Gated Events: 150147
Total Events: 150147

Gate	Events	% Gated
v CD45	143486	95.55
G2	1178	0.78
G3	1165	0.78
v CD34	1165	0.78
Beads	4681	3.12
all CD34	1167	0.78
lymphs	43821	29.19
all CD45	145423	96.85
Debris	0	0.00

viable CD34= 120.58 /ul

viable CD45= 14848.93 /ul

CD34 viability= 99.83 %

all CD34= 120.79 /ul

all CD45= 15051.48 /uL

% Viable CD45= 98.65

bead count = 48449.00

dilution factor 1.00

Sample volume 100.00 ul

viable CD34%= 0.81

Figure 4. Absolute viable CD34+ cell enumeration with the ISHAGE single platform protocol using TruCountTM Tube. Analysis of a fresh apheresis sample stained with CD45FITC, CD34PE and 7-AAD in a Trucount tube. After 20 minutes, sample lysed with ammonium chloride for 10 minutes at room temperature and listmode data acqiuired on FACSCalibur cytometer. A threshold was established on FL1 (CD45FITC) and an exclusion gate (R7) was established to prevent the inclusion of debris in the listmode file as described in the text. All Trucount beads were detected in region R6. Sample analysis was performed using Cellquest Pro software version 5.2 using semi-automated Editors.

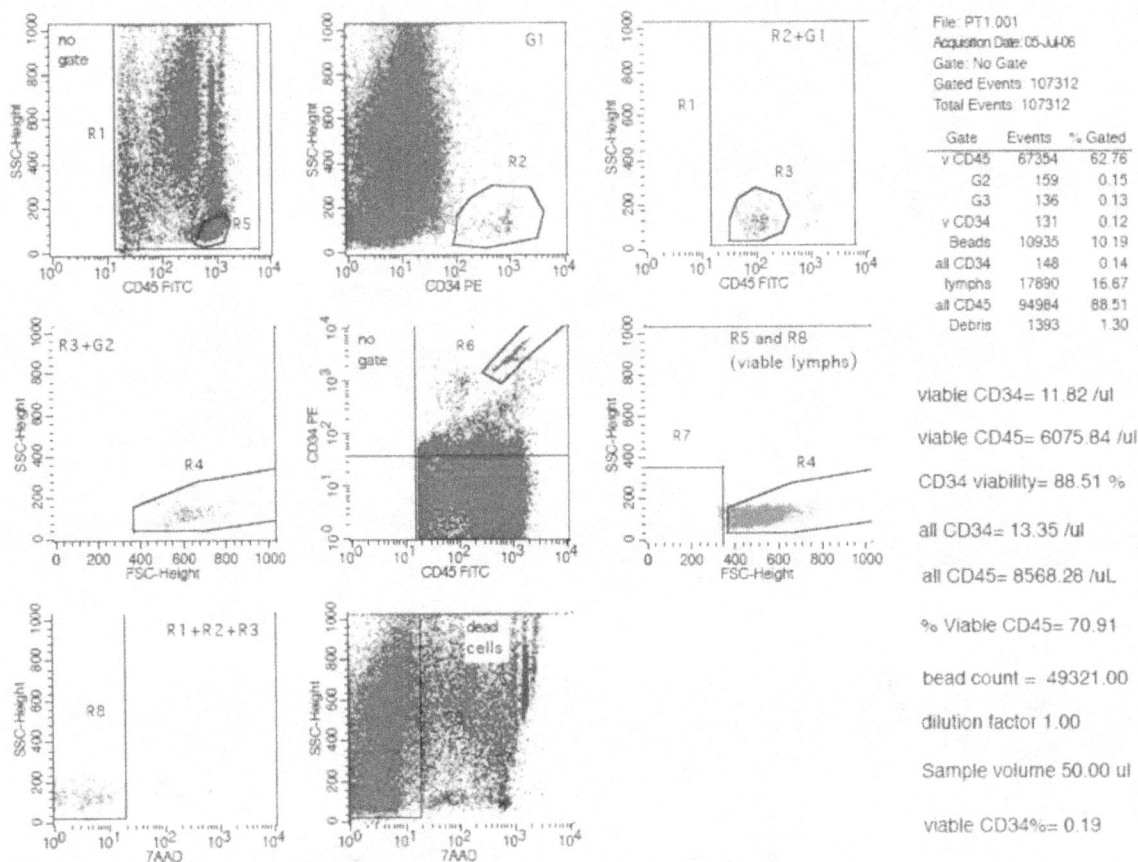

Figure 5. Absolute viable CD34+ cell enumeration with the ISHAGE single platform protocol using TruCountTM Tubes on a post-thawed cord blood sample. A frozen cord blood sample was thawed, processed as described and stained with CD45FITC, CD34PE and 7-AAD immediately post-thaw in a Trucount tube. After 15 minutes at room temperature, the sample was diluted with 1 ml of PBS/1%BSA (ammonium chloride lysis is not required nor recommended for post-thaw samples) and data acquired on FACSCalibur cytometer using Cellquest Pro software as described for Figure 4.

References

Brocklebank AM, Sparrow RL. Enumeration of CD34+ cells in cord blood: A variation on a single-platform flow cytometric method based on the ISHAGE gating strategy. Cytometry 2001 46:254-261.

Keeney M, Chin-Yee I, Weir K, Popma J, Nayar R, Sutherland DR. Single platform flow cytometric absolute CD34+ cell counts based on the ISHAGE guidelines. Cytometry 1998 34:61-70.

Keeney M, Sutherland DR. Current methods for identification of hematopoietic stem and progenitor cells in the clinical laboratory. In: Keren DF, McCoy JP Jr, Carey JL, eds. Flow cytometry in clinical diagnosis. 4th ed. Chicago: ASCP Press, 2007:1-24.

Sutherland DR, Anderson L, Keeney M, Nayar R, Chin-Yee I. The ISHAGE Guidelines for CD34+ cell determination by flow cytometry. J Hematother 1996 3:213-236.

Sutherland DR, Keating A, Nayar R, Anania S, Stewart AK. Sensitive detection and enumeration of CD34+ cells in peripheral and cord blood by flow cytometry. Exp Hematol 1994 22:1003-1010.

Sutherland DR, Nayyar R, Acton E, Giftakis A, Dean S, Mosiman VL. Comparison of two Single platform ISHAGE-based CD34 enumeration protocols on FACSCalibur[TM] and FACSCanto[TM] cytometers. Cytotherapy 2009 11:595-605.

16. Mesenchymal Stromal Cell Analysis

Vivek Tanavde
vivek@bii.a-star.edu.sg
Jyoti Kode
jkode@actrec.gov.in

Introduction

Identification of mesenchymal stromal cells (MSC) can be carried out using multiparametric flow cytometry. Bi-parametric dot plots of two cell surface markers can be used for defining regions to identify cell populations of interest. These regions can then be combined into gates for identifying MSC that satisfy all criteria of ISCT definition which states that "MSC are positive for CD73, CD105, CD90 and negative for CD34, CD45, HLA-DR, CD14 and CD19 expression" (Dominici 2006). The panel described in this protocol (Kode 2009; Kode and Tanavde, 2010) can be used to identify MSC on a flow cytometer with 488nm and 635nm excitation and four color detection.

Reagents

- CD73-PE (BD Pharmingen™, Catalog # 550257, clone-AD2, San Jose, CA)
- CD105-FITC (Abcam, Catalog # ab11415, clone SN6, Cambridge, UK)
- CD45-PerCP (BD Biosciences, Catalog # 347464, clone-2D1, San Jose, CA)
- CD90-APC (Thy1-APC) (BD Pharmingen, Catalog #559869, clone 5E10, San Jose, CA)
- CD34-APC (BD Pharmingen™, Catalog # 555824, clone-581, San Jose, CA)
- HLA-DR-PerCP (BD Biosciences, Catalog # 347364, clone-G46-6, San Jose, CA)
- CD14-APC (BD Biosciences, Catalog # 340436, clone-MϕP9, San Jose, CA)
- CD19-PerCp (BD Biosciences, Catalog # 347544, clone-4G7, San Jose, CA)
- Trypsin-EDTA (STEMCELL Technologies Inc, Catalog #07901, Vancouver, Canada)
- Phosphate Buffered Saline (PBS; 10mM, pH7.5)
- Wash buffer: PBS containing 1% FCS and 0.02% sodium azide

Method

1. Detach the MSC from flask surface using 0.25% Trypsin-0.04%EDTA.
2. Stop the trypsin action by washing with buffer.
3. Centrifuge the cells at 400 g for 5 minutes at 4^{0}C and resuspend in 0.5 ml wash buffer.
4. Count the cells using a hemocytometer or a cell counter.
5. Prepare the following panels of antibodies using antibody concentrations suggested by the manufacturer or determined by your titration against a known number of cells.
   Tube 1: Unstained cells. This tube serves as a control for measuring background fluorescence/

Tube 2: CD105-FITC/CD73-PE/CD45-PerCP/CD90-APC
Tube 3: CD105-FITC/CD73-PE/CD45-PerCP/CD34-APC
Tube 4: CD105-FITC/CD73-PE/HLA-DR-PerCP/CD90-APC
Tube 5: CD105-FITC/CD73-PE/CD19-PerCP/CD14-APC

6. Add 0.2×10^6 cells to each tube from the prepared cell suspension.
7. Incubate the tubes at 4^0C in dark for 30 minutes with intermittent tapping.
8. Add 1 ml of cold wash buffer to each tube. Mix cells and the reagent by tapping.
9. Centrifuge at 400 g for 5 minutes at 4^0C.
10. Aspirate the supernatant. Resuspend cells in 1 ml cold wash buffer and centrifuge.
11. Resuspend cells in 0.5 ml wash buffer for flow analysis.

Set up and Analysis

The protocols described here were carried out on a FACSCalibur (Becton Dickinson , San Jose, CA, USA) analyzer equipped with a 15 milliwatt 488 nm argon ion laser and a 635 nm red diode laser. The following emission filters were used: 530 nm (FITC), 585 nm (PE/PI), and >670 nm (PerCP) with base unit, and 661 nm (APC) with FL4 option. Of these fluorochromes, APC is excited with the 635 nm red diode laser whereas the other 3 fluorochromes are excited with the 488 nm argon laser. However, any comparable instrument can be used for this analysis. Single stained controls or compensation beads can be used for setting instrument compensation. Instead of unstained cells, fluorescence minus one (FMO) controls are also useful for setting thresholds above which the signal will be considered positive. The following schema based on the expression of different markers will be used for identifying true MSC:

Tube 2: This tube identifies cells which are CD105+/CD73+/ CD45-/ CD90+.
Tube 3: The CD 105+/CD73+/CD45- population identified in this tube should be negative for CD34.
Tube 4: The CD105+/CD73+/ CD90+ population in this tube should be negative for HLA-DR.
Tube 5: The CD105+/CD73+ population in this tube should be negative for CD19 & CD14.

Thus the CD105+/CD73+ cells will be CD90+ and negative for CD45, CD34, HLA-DR, CD19 and CD14 expression. As these cells satisfy the ISCT definition, they can be considered to be the true MSC.

Comments

- MSC are large cells (>30 μM in diameter). Therefore the settings on the FSC detector have to be adjusted so that the cells are displayed within the FSC linear scale.

- MSC like fibroblasts are naturally adherent and tend to clump. This can pose problems in sample acquisition since flow cytometry requires single cells for accurate enumeration. Cell clumping can be avoided by having EDTA in the staining buffer and straining the cells through a 40 μm strainer just before flow analysis. Extensive cell clumping can be avoided by addition of 10 ng/ml DNase1 (Sigma Corporation, USA) to the staining solution.

References

Dominici M, Le Blanc K, Mueller I, Slaper-Cortenbach I, Marini F, Krause D, Deans R, Keating A, Prockop D, Horwitz E. Minimal criteria for defining multipotent mesenchymal stromal cells. The International Society for Cellular Therapy position statement. Cytotherapy 2006 8:315-317.

Kode J, Mukherjee S, Joglekar M, Hardikar A. Mesenchymal Stem cells: Immunobiology and role in Immunomodulation and Tissue regeneration. Cytotherapy, 2009 11:377-391.

Kode J, Tanavde V. Mesenchymal Stromal Cells and Their clinical applications. In: Applications of Flow Cytometry in 'Stem Cell Research and Tissue Regeneration', Eds. Awtar Krishan, H Krishnamurthy and Satish Totey, Wiley- Blackwell Publications. July 2010 (In Press).

17. Flow Cytometric Analysis of Spermatogonial Stem Cells

BS Srinag
srinag@ncbs.res.in
JM Kalappurakkal
joseph@ncbs.res.in
H Krishnamurthy
krishna@ncbs.res.in

Introduction
Spermatogenesis is a complex process that involves a series of divisions and differentiation events by which; a Spermatogonial Stem Cell (SSC) is progressively transformed into mature sperm or male gametes. SSCs are found in low frequency and persist throughout the lifetime of the male and are involved in tissue homeostasis.

The SSCs arise from the Primordial germ cells (PGCs) or gonocytes. The PGCs differentiate into A_{single} (A_s) spermatogonia which are considered to be the SSCs. The SSCs can undergo both self-renewal and differentiation. The spermatogonial cells undergo meiosis to form the primary and secondary spermatocytes, which give rise to haploid spermatids resulting in the ultimate formation of mature spermatozoa.

Purification of SSC is important for several reasons. Studying the mechanisms that control SSCs, their proliferation and differentiation may be important for the treatment of male infertility and to understand the etiology of testicular tumors. SSCs of patients undergoing radiotherapy or chemotherapy can be purified, expanded, stored and subsequently transplanted to restore fertility. The cultured SSCs which are a source of adult stem cells could be used to treat damaged tissues.

Techniques used in the analysis of SSCs
The A_s (including SSCs) A_{paired} (A_{pr}) and $A_{aligned}$ (A_{al}) germ cells, collectively referred to as the undifferentiated spermatogonia, share similar phenotypic and probably similar molecular characteristics. However, at present, there is no phenotypic, biochemical, or molecular markers to differentiate between these populations. Neurogenin 3 is specifically expressed in the undifferentiated spermatogonial cells (Yoshida et al., 2004). The differentiating spermatogonia are distinguishable from the undifferentiated spermatogonia on the basis of several phenotypic characteristics. By flow cytometric analysis, SSC are recognized with low side scatter and expression of the following specific surface antigens: CD49f[+] (Shinohara et al., 1999), β1-integrin[+] (Shinohara et al., 1999), CD9[+] (Kanatsu-Shinohara et al., 2004), Thy-1[+] (Kubota et al., 2004), c-kit[-] (Ohta et al., 2003), GGfrα1[+] (Buageaw et al., 2005; Ebata et al., 2005) and Ep-CAM. Side population (SP) analysis, which is based on the vital dye efflux properties of certain primitive cell populations, has been used for the identification of stem cell populations in a variety of tissues. Testicular cells also contain a distinct population of cells with SP phenotype and express SSC markers like α6-integrin and Stra8 (Lassalle et al., 2004).

Materials and reagents required for SP analysis:
- Mouse testis
- Dissection kit: 1 surgical blade/ scalpel, 1 blunt forceps, 2 sharp forceps, 1 sharp scissors.
- Tissue culture 60 mm Petri dishes
- Pipettes, tips (1000 µl and 200 µl), 15 ml Falcon tubes and flow analyzer sample (FACS) tubes.
- Cell strainer or 40 µm nylon mesh
- Calcium, magnesium free Hanks' Balanced salt solution (HBSS) pH 7.4
- Trypsin–EDTA solution – 1 mM ethylene diamine tetra acetic acid (EDTA), 0.25% trypsin in HBSS
- Collagenase type I (1 mg/ml) in HBSS
- DNase I solution – (Sigma) 7 mg/ml in HBSS
- Fetal bovine serum (FBS)
- Hoechst 33342 – Stock concentration 5mg/ml in DMSO, stored at -20°C and working concentration of 1mg/ml in DMSO, stored at room temperature.
- Propidium iodide: Stock concentration of 5 mg/ml in HBSS and stored at -20ºC. Working concentration of 1 mg/ml is prepared in HBSS and stored at 4°C.
- Verapamil- Stock concentration 2.5 mg/ml in water, freshly prepared in dark.
- Incubation buffer- 20 mM HEPES, 1.2 mM $MgSO_4 \cdot 7H_2O$, 1.3 mM $CaCl_2 \cdot 2H_2O$, 6.6 mM sodium pyruvate , 0.05% lactate, glutamine, pH7.2 , 1% serum in HBSS.
- Trypan blue solution for viability counting: 0.4% in PBS.

Preparation of Testicular Cells (Enzymatic method)

The protocol for preparing single cell suspension of the testicular cells which can later be processed for isolating SSCs by flow or magnetic cell sorting is described below (Oatley and Brinster, 2006).

6-8 day old mouse pups produce the most enriched population of stem cells. For a typical preparation, 8–10 pups are used.

1. Anaesthetize the mouse pups (6-8 days old) with ether and perform a cervical dislocation. Place the mice dorso-ventrally flat and make a small incision in the lower abdominal region near the scrotum and remove the testis.
2. Place the testes in sterile calcium and magnesium free HBSS and remove the tunica albuginea surgically.
3. Transfer the tissue into a tube containing 4.5 ml of trypsin–EDTA solution or 2 ml of Collagenase solution and 0.5 ml of DNase I solution.
4. Disperse the tubules gently and incubate in shaking water bath at 37°C for 5-10 min followed by gentle pipetting and addition of 0.5 ml of DNase I solution.
5. Continue digestion at 37°C for another 5 min followed by addition of 1 ml of FBS.
6. Filter the cell suspension through a 40 um cell strainer; wash with HBSS and the pellet the cells by centrifugation at 600 g for 8 min at 4°C. Re-suspend the cell pellet in HBSS until further use.
7. Count the viable cells using trypan blue staining.

Side population (SP) analysis of SSCs by flow cytometry:

The cells which actively efflux DNA binding dyes (Hoechst 33342) have a characteristic Side Population (SP) profile with low Hoechst fluorescence. This population is highly enriched in stem cells (Lassalle et al., 2004).

Staining protocol

This staining protocol is according to Lassalle et al. (2004).

1. Prepare the single cell suspension of testis by the enzymatic digestion and rinse cells in incubation buffer and incubate at 32°C.
2. Incubate 10^6 cells per ml in incubation buffer with Hoechst 33342 (5 µg/ml) for 1 hr at 32°C, with gentle stirring every 10 min.
3. After incubation, pellet the cells by centrifugation at 600 g for 5 min at 4°C and wash twice with ice cold incubation buffer. Prior to analysis, add propidium iodide (2 µg/ml) to stain and identify dead cells by flow analysis. Keep the tubes on ice until further analysis.
4. For control, pre-incubate the cells with blockers of ABCG2 transporters like Verapamil (30 µl/ml of stock to get a final concentration of 75 µg/ml) and incubate for 30 min at 32°C and stain the cells with Hoechst 33342 in the presence of inhibitor as previously described. Use RBC-lysed bone marrow cells as internal control for SP analysis of Hematopoietic Stem Cells (Goodell et al., 1996).

Data acquisition

The flow cytometer should be equipped with a 488 nm and 355/375 nm lasers. The emission filter combinations required for collection of the blue and red emission from cells stained with Hoechst 33342 are 450/50 nm BP and 650 nm LP, respectively. The PI emission filter should be a 585/15 band pass and a 580 nm LP dichroic beam splitter.

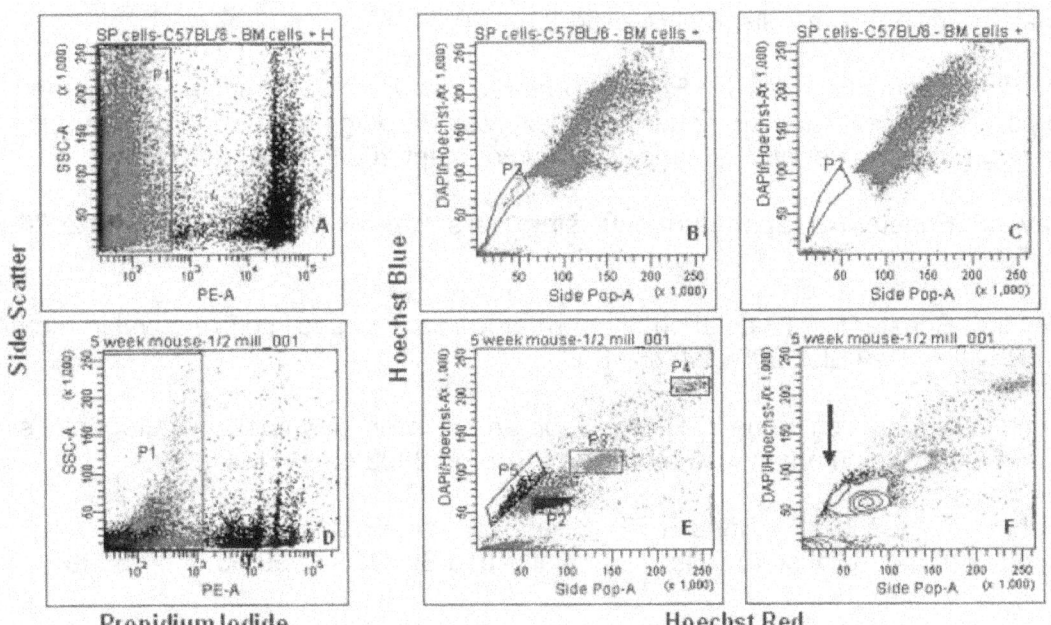

Data Analysis and interpretation

As shown in Figure 1A and 1D, side scatter vs. propidium iodide fluorescence is used to exclude the dead cells and gate on the live cells (gate P1). Figures 1B show the presence of the SP cells (gate P2) in the bone marrow sample. These cells are absent in samples incubated with Hoechst 33342 and Verapamil (Figure 1C).

The PI negative testicular cells in gate P1 of figure 1D, contained four major populations of haploid (P2), diploid (P3) and tetraploid (P4) cells in figure 1E. In addition, the testicular cells show the presence of SP cells (gate P5 in figure 1E) and the same population is clearly seen in the contour plot Figure 1F (arrow).

References

Buageaw A, Sukhwani M, Ben-Yehudah A, Ehmcke J, Rawe VY, Pholpramool C, Orwig KE, Schlatt S. GDNF family receptor α1 phenotype of spermatogonial stem cells in immature mouse testes. Biol. Reprod 2005 73:1011–16.

Goodell M A, Brose K, Paradis G, Conner AS, Mulligan RC. Isolation and functional properties of murine hematopoietic stem cells that are replicating in vivo. J Exp Med 1996 183:1797-806.

Kubota H, Avarbock MR, Brinster RL. Culture conditions and single growth factors affect fate determination of mouse spermatogonial stem cells. Biol Reprod 2004 71:722–731.

Kubota H, Avarbock MR, Brinster RL. Growth factors essential for selfrenewal and expansion of mouse spermatogonial stem cells. Proc Natl Acad Sci 2004 101:16489–16494.

Kanatsu-Shinohara M, Toyokuni S, Shinohara T. CD9 is a surface marker on mouse and rat male germline stem cells. Biol Reprod 2004 70:70–75.

Lassalle B, Bastos H, Louis JP, Riou L, Testart J, Dutrillaux B, Fouchet P, Allemand I. Side population cells in adult mouse testis express Bcrp1 gene and are enriched in spermatogonia and germinal stem cells. Development 2004 131:479-487.

Oatley JM, Brinster RL. Spermatogonial stem cells. Review Methods Enzymol 2006 419:259-82.

Oatley JM, Brinster RL. Regulation of spermatogonial stem cell self-renewal in mammals. Annu Rev Cell Dev Biol. 2008 24:263-86.

Ohta H, Tohda A, Nishimune Y. Proliferation and differentiation of spermatogonial stem cells in the W/Wv mutant mouse testis. Biol Reprod 2003 69:1815–21.

Shinohara T, Avarbock M, Brinster RL. β1 and α6 integrin are surface markers on mouse spermatogonial stem cells . Proc Natl Acad Sci USA 1999 96:5504-5509.

18. Tumor Stem Cell Marker Expression

Awtar Krishan
akrishan@med.miami.edu.
Ronald M. Hamelik
rhamelik@med.miami.edu

Introduction

Several recent studies have focused on the presence and importance of tumor stem cells in human solid tumors (Al-Hajj et al., Ginestier et al.). Expression of cell surface markers (e.g., CD44, CD133), certain detoxifying enzymes such as ALDH1 and drug efflux characteristics (side population or SP phenotype) have been used to identify and sort tumor stem cells (Krishan et al., O'Brien et al., Wright et al.).

As CD44$^+$/CD24$^-$ expression is also seen in certain ALDH1^{bright} non-epithelial hematopoietic cells present in body cavity fluids, we have used the following protocols for monitoring the expression of Aldehyde dehydrogenase 1, Ber-EP4 (same as Epithelial Surface Antigen or ESA), CD44 and CD24 in epithelial cells from body cavity fluids of female patients (Krishan et al.).

Reagents

- Phospate buffered saline (PBS): (Gibco 14190) Dulbecco's PBS
- 0.4% Trypan blue: (Gibco 15250-061)
- Histopaque-1077: (Sigma 10771)
- Propidium iodide in hypotonic sodium citrate:
    5 mg Propidium iodide (Sigma P-4170)
    100 mg Sodium citrate (Sigma S-4641, trisodium dehydrate)
    100 ml glass distilled water
    30 µl Nonidet P-40 non-ionic detergent. Since NP-40 may be difficult to locate, one can use Igepal CA-630 (Sigma I-3021) instead.
- Ber-EP4-FITC: (Dako F0860, clone Ber-EP4)
- CD44-RPE: (BD Pharmingen 555479, clone-G44-26)
- CD24-PE-Alexa Fluor® 610 conjugate: (Invitrogen MH CD2422, clone SN3)
- Aldefluor® reagent kit: (StemCell Technology 01700)
- Biotin conjugated Epithelial Specific Antigen: (GeneTex GTX72682, clone B29.1-VU-ID9)
- Peridinin chlorophyll-a protein (PerCP) conjugated strepavidin: (BD Biosciences 341130)

Method
Processing of body cavity fluids

1. After coarse filtration thru gauze, prepare cell pellets by centrifugation at Relative Centrifugal Force (RCF) x 220 for 5 min at 4°C.
2. Wash once with 5 ml of PBS.

3. Samples with a large number of red blood cells (RBCs), should be diluted in PBS, layered over Histopaque and centrifuged at 400 x RCF for 30 min at room temperature.
4. The cells at the interface should be carefully aspirated and washed twice with cold PBS.
5. Determine the number of viable cells by trypan blue or propidium iodide viability assays.
6. Stain an aliquot of $2x10^6$ cells with hypotonic propidium iodide for flow cytometric analysis of DNA content and aneuploidy.

Ber-EP4/CD44/CD24 Staining
1. Wash an aliquot of $2x10^6$ cells with 3 ml of PBS, centrifuge at 220 x RCF for 5 min at 4°C and decant.
2. Resuspend the cell pellet and add 10 µl of Ber-EP4-FITC, 20 µl of CD44-RPE and 5 µl of CD24-PE-Alexa Fluor® 610 conjugate. Mix well.
3. Incubate for 20 minutes at room temperature in the dark.
4. Add 3 ml of PBS, centrifuge 220 x RCF for 5 min at 4°C and decant.
5. Resuspend the cell pellet in 2 ml of PBS for flow cytometric analysis.

Aldefluor®/CD44/CD24/ESA Staining:
1. Centrifuge an aliquot of $2x10^6$ cells and wash with 3 ml of PBS.
2. Resuspend the cell pellet in 1 ml of Aldefluor® buffer containing 5 µl of activated Aldefluor® reagent.
3. Transfer one half of the sample to a second tube containing 5 µl of the ALDH1 inhibitor, diethylaminobenzaldehyde (DEAB).
4. Mix and incubate for 60 min in a 37°C water bath with intermittent shaking.
5. Remove from the water bath and add 20 µl of CD44-RPE, 5 µl of CD24-PE-Alexa Fluor® 610 and 3 µl of biotin conjugated ESA antibodies.
6. Mix well and incubate for 20 min at room temperature in the dark.
7. Add 3 ml of PBS, centrifuge 220 x RCF for 5 min at 4°C and decant.
8. Add 10 µl of PerCP conjugated strepavidin.
9. Incubate for 20 min at room temperature in the dark.
10. Add 3 ml of PBS, centrifuge 220 x RCF for 5 min at 4°C and decant.
11. Resuspend the pellet in 400 µl of Aldefluor® buffer for flow analysis.

Flow Cytometry
1. Set threshold and electronic gates in Forward Scatter versus Side Scatter plots to exclude debris and any residual red blood cells.
2. Collect fluorescent signals from Aldefluor®, CD44-RPE, CD24-PE-Alexa Fluor® 610, and Ber-EP4 PerCP signals using 525/40, 575/30, 620/30 and 675/40 band pass filters respectively.
3. Use electronic gates to identify cells with ALDH1 bright expression by comparing dot plots of cells incubated with Aldefluor® with or without the ALDH1 inhibitor, DEAB.
4. Analyze Ber-EP4 pos and ALDH1 bright cells for CD44 and CD24 expression.

5. Perform list mode color compensation and data analysis using WinList 3D software (Version 5.0, Verity Software, Inc, Topsham, ME) and/or WinMDI (Version 2.9, The Scripps Research Institute, La Jolla, CA).

Results

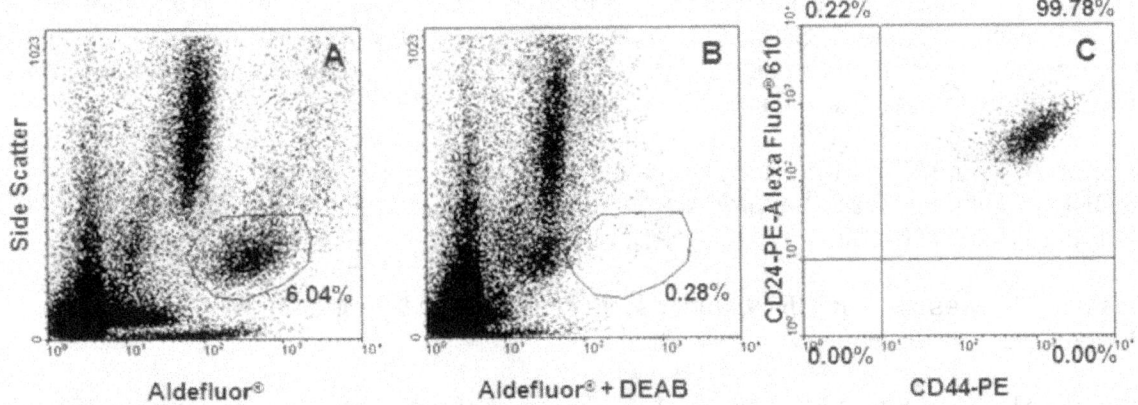

Figure 1. DNA histogram 1A shows that a large number of cells in this sample were aneuploid with DNA index of 1.86. Dot plot 1B of Forward Scatter vs. Ber-EP4 expression shows that a large and distinct population (66.66%) of the cells had Ber-EP4^{pos} expression. Dot plot 1C of Ber-EP4^{pos} cells shows that 50.10% and 7.08% of these cells had CD44$^+$/CD24$^+$ and CD44$^+$/CD24$^-$ expression, respectively.

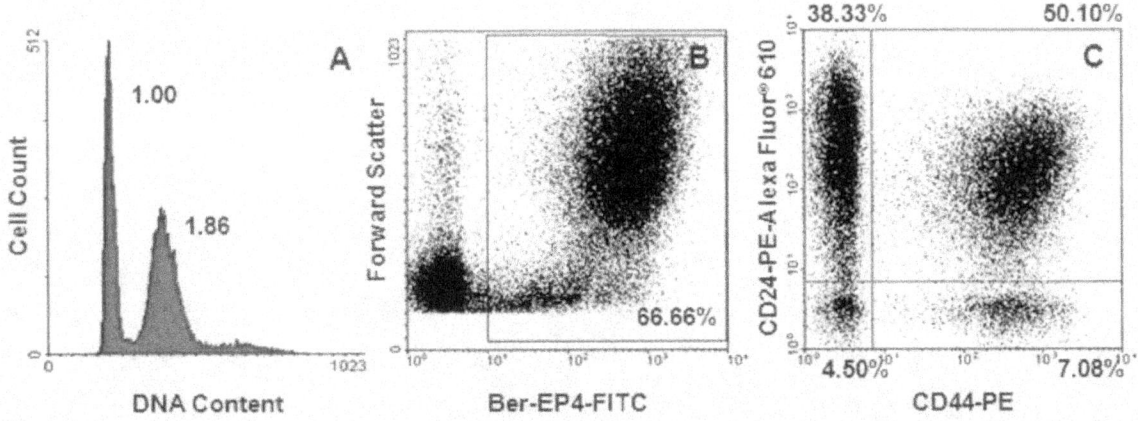

Figure 2 shows cells from a pleural fluid. A comparison of dot plots 2A and 2B show the presence of cells (6.04%) with ALDH1^{bright} content (gate, Figure 2A). Analysis of Ber-EP4 expression showed that 72% of the ALDH1^{bright} cells had ESA^{pos} expression (data not shown). Dot plot 2C shows that almost all of the ALDH1^{bright}/ESA^{pos} cells had CD44$^+$/CD24$^+$ expression and none were CD44$^+$/CD24$^-$.

References

Al-Hajj M, Wicha MS, Benito-Hernandez A, Morrison SJ, Clarke MF. Prospective identification of tumorigenic breast cancer cells. Proc Natl Acad Sci USA 2003 100:3983-3988.

Ginestier C, Hur MH, Charafe-Jauffret E, Monville F, Dutcher J, Brown M, Jacquemier J, Viens P, Kleer CG, Liu S, Schott A, Hayes D, Birnbaum D, Wicha MS, Dontu G. ALDH1 is a marker of normal and malignant human mammary stem cells and a predictor of poor clinical outcome. Cell Stem Cell 2007 1:555-567.

Krishan A, Sharma D, Sharma S, Hamelik RM, Ganjei-Azar P, Nadji M. ALDH(+)/CD44(+)/CD24(-) expression in cells from body cavity fluids. Cytometry B Clin Cytom 2010 78:176-182.

O'Brien CA, Kreso A, Dick JE. Cancer stem cells in solid tumors: an overview. Semin Radiat Oncol 2009 19:71-77.

Wright MH, Calcagno AM, Salcido CD, Carlson MD, Ambudkar SV, Varticovski L. Brca1 breast tumors contain distinct CD44+/CD24- and CD133+ cells with cancer stem cell characteristics. Breast Cancer Res 2008 10:105.

19. Flow Cytogenetic Analysis of Single Chromosomes (One color analysis)

L. Scott Cram
scottcram42@gmail.com
Claire Sanders
csanders@lanl.gov

Introduction
Chromosome analysis and sorting (Flow Cytogenetics) was first described in 1978. This technology provided the seminal technology for initiation of the human genome project. For the first time one could obtain large numbers of highly purified human chromosomes of each type. From each of the 24 different types of flow sorted human chromosomes, large insert chromosome libraries were constructed and the DNA sequenced. More recently the technology has been applied to the construction of chromosome specific probes from a large variety of species and for clinical applications involving sorting of complex karyotypes.

Basic Technique
For many species each chromosome type can be resolved using flow cytometric techniques. Once resolved, they can be sorted with high purity. There are two critical steps to achieve high resolution and high sorting purity; chromosome isolation and staining, and instrument alignment. The procedure presented here is recommended for anyone getting into flow cytogenetics for the first time. It provides a straight forward isolation and staining procedure for isolating large numbers of stabilized chromosomes that can be used for instrument optimization. Once this basic technique is mastered most teams feel comfortable proceeding to two color analysis and chromosome sorting.

Preparation and Analysis of Chinese Hamster Chromosomes
Euploid Chinese hamster cells (not CHO cells) can be prepared from Chinese hamster embryos, cultured for a few (5) passages, cloned, and carried in culture being careful to maintain monolayer cultures. These cell strains will become aneuploid after 20-30 passages. Once they become aneuploid their karyotype will become increasingly difficult to resolve because of the large number of chromosomal rearrangements.

Typically referred to as the "Hypotonic Propidium Iodide" protocol, there are six basic steps:
1. Arrest cells in metaphase with Colcemid.
2. Collect mitotic cells by mitotic shake off procedure.
3. Swell the cells with hypotonic KCl.
4. Rupture the cell membrane with detergent and disperse chromosomes by forcing suspension through a syringe needle.
5. Stain the chromosome with DNA specific dye(s).
6. Analyze preparations.

Reagents

- Colcemid stock solution: 1 mg/ml in Balanced Salt Solution.
- Hoechst 33342 stock solution.
- Reagent 1 - Hypotonic Propidium Iodide solution: 50 µg/ml PI in 75 mM KCl solution.
- Reagent 2 - Hypotonic PI Triton X-100 RNase solution: 50 µg/ml PI, 1% Triton X-100, 1 mg/ml RNase in 75 mM KCl solution.

Method

1. Colcemid blocks a sub-confluent and exponentially growing cells in mitosis by depolymerizing the spindle microtubules. The concentration sufficient to block cells in mitosis varies between 0.01 µg/ml to 0.1 µg/ml depending on different cell types. Cells with MDR phenotype will need co-incubation with an efflux blocker to retain intracellular Colcemid concentration. We generally use the higher concentration for our cloned Chinese hamster cells. One to two T-150 flasks that are 50 -60% confluent yield enough cells for flow analysis.
2. Incubate cells in Colcemid for 3 hours, 37°C. Gently but firmly shake off mitotic cells by slapping the flask and recover by centrifugation (100xg for 6 min). This is a critical step and can be monitored in any suitable tissue culture scope.
3. OPTION. Prior to centrifugation, remove a small amount of media and stain with 10 µM Hoechst 33342 to determine the total number of cells and the percentage of mitotic cells.
4. After centrifugation, remove ALL tissue culture media and gently resuspend the pellet.
5. Add 0.5 ml of Reagent 1, the hypotonic PI KCl solution.
6. Allow the cells to swell for 10 minutes at room temperature.
7. Monitor cells under a microscope. Ideal conditions result in mostly intact cells with intact membranes and are therefore, non-fluorescent. Cells will soon become fluorescent as noted in the paragraph below: "Mitotic cell swelling".
8. To 0.5 ml of cells, add 0.25 ml of Reagent 2, the hypotonic PI detergent KCl solution with RNase.
9. Incubate 3 min at room temperature to partially dissolve cell membranes.
10. Forcefully syringe the cell suspension 4-5 times through a 1.5 inch, 22 gauge needle with the point of the needle pointing against the wall of the tube.
11. Incubate the chromosome suspension for 30 min at 37°C and filter thru a 60 µm nylon mesh.
12. The isolated chromosomes are ready for analysis and can be maintained for at least several days if stored at 4°C,

Tips and Hints

- **Culture conditions:** Exponential cell growth is essential for optimum results. This requires maintaining log phase cultures at sub confluent condition for a couple of passages. Monolayer cultures allow one to easily establish and monitor growth rates and conditions, in particular single cells with minimal cell to cell contact inhibition. Rapidly growing fibroblasts cultures can be split at one to five or one to ten and re-cultured before they become confluent. For optimum results, <u>fresh</u> culture media can be added just prior to the shake off procedure. This step removes any floating and/or dead cells or debris just prior to recovery of the mitotic cells.

- **Number of cells:** Cell concentration is critical. At least 10^6 mitotic cells from each of two T-150 flasks are collected as noted above in step #2. Lower cell concentrations have consistently resulted in chromosome suspensions of lower quality.

- **Mitotic cell swelling:** Probably the most critical step in the procedure and one that can be monitored as noted in step #3, above. The presence of Hoechst 33342 in the culture media yields fluorescent viable cells. When observed under a fluorescent microscope (step #7) mitotic cells will have a blue nucleus (Hoechst staining) and no red intracellular staining (propidium iodide (PI) <u>provided</u> the cell membrane remains intact. As the cells swell and their membrane first becomes permeable, one can observe the diffusion of PI into the cell. The diffusion of PI across a cell will result in a gradual transition (few seconds) from blue to red chromosomes. This is the best method of monitoring cell swelling. It's also a dramatic event to observe.

- **Sterile reagents:** If one wishes to hold prepared samples for an extended period of time (generally greater than two weeks), the use of sterile reagents and careful sample handling will provide samples that have sometimes been reanalyzed one to two months after isolation. Prepared samples are stored at 4°C.

- **Colcemid concentration:** Colcemid concentration (step #1) varies with the cell type. Final concentration can be determined by observing mitotic cells (stained with Hoechst) under a microscope. Too low a concentration will result in too few cells blocked in mitosis and/or cells escaping mitotic block. Too high a concentration will result in hyper condensed nuclei that appear distorted.

- **Hypotonic conditions:** Hypotonic concentrations between 50 and 75 mM KCl are sufficient to cause osmotic lysis of most fibroblast monolayer cultures. Monitoring cell swelling (step #7) can be used to determine optimum hypotonicity.

- **Flow cytometer filters:** Emission filters should be carefully chosen. Fluorescence emission of PI is centered around 620 nm and is very wide. Because the chromosomes are small it is important to collect the maximum light possible. To do this, a long pass filter around 550 nm is the best choice.

Artifacts

- Broken or severely stretched chromosomes sometimes appear when preparations are observed under the microscope. This may be due to sheer forces generated when the cover slip is placed on the slide. Generally these artifacts due not present a problem. Most often broken/stretched chromosomes will appear to have two 'normal' arms connected by a fine filament of DNA. After having prepared literally a few hundred samples, we have concluded that their appearance in a microscope is not indicative of chromosome resolution measured with a flow cytometer.

- The procedure does not include a step to remove extra cellular debris generated in the isolation of mitotic chromosomes. Such debris is non-fluorescent and is not analyzed by the flow cytometer, so it does not interfere with flow cytometric results. However, it can contribute to light scatter, rendering the scatter parameter less useful for triggering than fluorescence.

Data Interpretation

Data interpretation of high quality well resolved peaks is straightforward. The area under a peak is the number of chromosomes of a particular type. For euploid cells, the area under each peak will be the same except for X, Y individuals. In such cases the peak resulting from the X and Y chromosomes will have half the events as in the autosomal peaks. Structural and numerical aberrations will alter, respectively, the appearance of the peaks and the area under a peak. Peak position (channel number) is proportional to DNA content for PI staining.

References

Cram LS, Bartholdi MF, Ray FA, Cassidy MC, Kraemer PM. Univariate Flow Karyotype Analysis, In: Flow Cytogenetics (Ed. Gray, J. W.) Academic Press pp 113-136 (1989).

Cram LS, Bell CS, Fawcett JJ. Chromosome Sorting and Genomics, Methods in Cell Science 2002 24:27-35.

Cram LS, Gray JW, Carter NP. Cytometry and Genetics, Cytometry 2004 58:33-36.

Cram LS, Ray FA, Bartholdi MF. Univariate Analysis of Metaphase Chromosomes Using the Hypotonic Potassium Chloride-Propidium Iodide Protocol. In: Methods in Cell Biology 1990 33:369-377.

20. Flow Cytometric Analysis of Testicular Cells

Navya Jain
navya@ncbs.res.in
B.S. Srinag
srinag@ncbs.res.in
H. Krishnamurthy
krishna@ncbs.res.in

Introduction

Cell division involves a series of events which eventually lead to the division and duplication of the genetic material. The process of cell division is of two types: Mitosis (equational division) and meiosis (reduction division). Mitosis results in formation of two diploid daughter cells whereas in meiosis a diploid cell will terminally differentiate into haploid gametes. In mammalian spermatogenesis, a diploid spermatoginal stem cell apart from self renewing itself, will undergo a series of divisions and terminally differentiates into a haploid sperms.

Flow cytometric method to quantify testicular cells

Propidium iodide, Hoechst, DAPI, acridine orange, 7-aminoactinomycin D (7-AAD), chromomycin A3 are some of the fluorescent dyes used for flow cytometric analysis of DNA content of cells

Propidum iodide (PI) is a non-permeant, nucleic acid intercalating dye that penetrates the membranes of dying/dead cells. PI binds to the major and the minor grooves of the DNA. It can be excited with the 488 nm laser line and emits in the range of 550 to 750 nm with an emission maxima at 620 nm.

Materials and Regents

- Dissection kit: 1 surgical blade/scalpel, 1 blunt forceps, 2 sharp forceps, 1 sharp scissors.
- Tissue culture 60 mm Petri dishes.
- Pipettes, tips (1000 µl and 200 µl), 15 ml Falcon tubes.
- Cell strainer or 40 µm nylon mesh.
- Calcium, magnesium free Phosphate Buffered Saline (PBS), pH 7.4.
- Chilled absolute ethanol.
- Propidium iodide: Stock concentration of 5 mg/ml in PBS and stored at -20°C.
- RNAse Type III: Stock concentration of 10 mg/ml in PBS and stored at -20°C.
- Pepsin: Working concentration of 0.25% of pepsin in PBS; pH adjusted to 2.0 with 1N HCl and stored at -20°C.
- NP 40 or IGEPAL CA 450 detergent
- Propidium iodide staining solution: 25 µg PI, 40 µg RNAse, 0.03% NP 40 or IGEPAL CA 450 in PBS.

Preparation of cells
1. Anaesthetize the mice/rat (8-weeks old) with ether and perform a cervical dislocation or use a method cleared by the Institutional Animal care and Ethics committee.
2. Make a small incision in the lower abdominal region near the scrotum and remove the testis.
3. Place a testis in 1 ml of sterile calcium, magnesium free 1X PBS in 60 mm dishes and remove the tunica albuginea.
4. Gently tease the tissue mass with a pair of scissors followed by gentle pipetting.
5. Filter the cell suspension through a 40 μm cell strainer or nylon mesh.
6. Make up the volume of cell suspension up to 3 ml with PBS.

Fixation
1. Fix the cells in suspension, by adding 7 ml of chilled absolute ethanol. Mix and label the tubes.
2. Leave the fixed cells at 4°C for at least for 24 hrs before flow cytometric analysis.

Staining of cells
1. Centrifuge an aliquot of fixed cells at 600 g for 5 mins at 4°C.
2. Wash the pellet with 1X PBS (twice) and spin down the cells at 600 g for 5 min at 4°C to ensure the complete removal of residual ethanol.
3. Add 1 ml of pepsin solution and incubate for 10 min at 37°C, pellet the cells at 600 g for 5 min at 4°C.
4. Wash the pellet with 1X PBS and spin down the cells at 600 g to ensure the complete removal of pepsin.
5. Incubate the cells with 1 ml of PI staining solution for 15 min before flow cytometric analysis.

Tips:
- Pepsin preparation: Weigh the required amount of pepsin into appropriate container and add cold 1X PBS. Add 1N HCl drop wise with constant stirring and adjust the pH to 2. Ensure that the solution does not turn milky upon the addition of HCl.
- Ethanol fixation of cells: Add ethanol drop wise to the tube containing cells with constant shaking or vortexing to avoid cell clumping.

Data acquisition and analysis
The flow cytometer should be equipped with 488 nm laser, an emission filter (585/15 nm band pass) to collect PI signals and doublet discrimination module. Open FSC vs. SSC dot plot, PI-Area vs. PI-Width plot and PI-Area histogram. Run the PI stained testicular cells and set the PMT voltage of PI channel in such a way that all four populations are in the PI-Area histogram window (Figure 1B). Acquire 10000 cells and store the data as a list mode data file. To analyze the data, set the region around the singlet population on the PI-Area vs. PI-Width plot as shown in Figure 1A and apply this region to PI-Area histogram (Figure 1B).

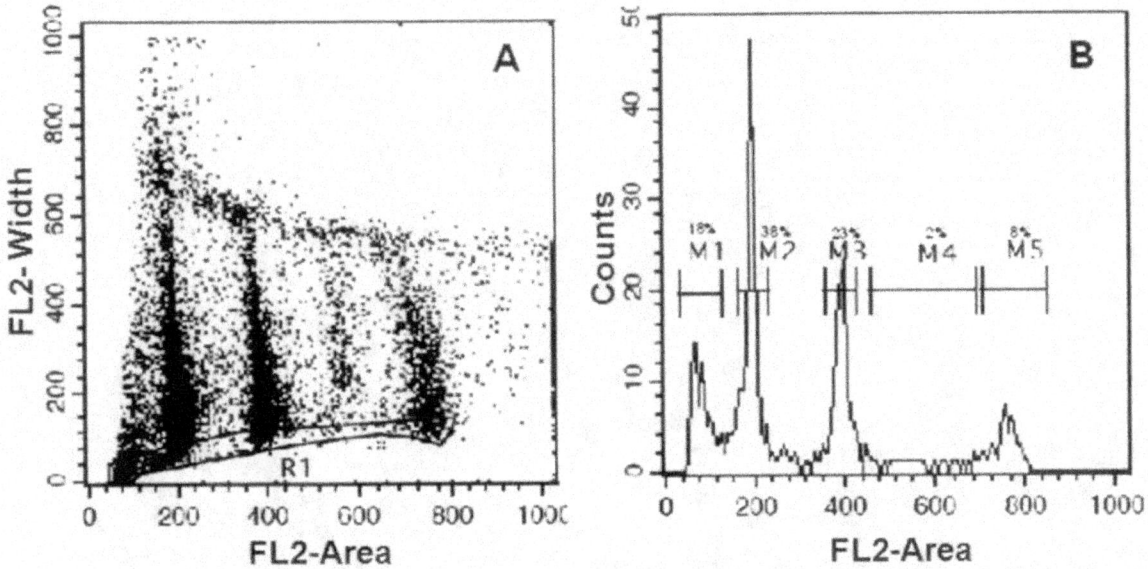

Figure 1B shows DNA distribution histograms of PI-stained testicular cells. Five distinct populations marked as M1-M5 are shown. Based on their DNA content, the five major populations are of elongated spermatids (M1), round spermatids (M2); spermatogonia and testicular somatic cells (M3); spermatogonial cells in DNA synthesis (M4), primary spermatocytes and G_2 spermatogonia (M5). It may be noted that the percentage of testicular somatic cells (Sertoli, Leydig, and peritubular myoid cells) is less than 3% of total testicular cells and falls within the 2C population. Set the marker for each population as shown in Figure 1B to get the percentage of each population.

References

Krishnamurthy H, Babu PS, Morales CR and Sairam MR. Delay in sexual maturity of the follicle-stimulating homone receptor knockout mouse (FORKO). Biology of Reproduction 2001 65:522-531.

Krishnamurthy H, Danilovich N, Morales CR, Sairam MR. Qualitative and quantitative decline in spermatogenesis of the follicle-stimulating hormone receptor knockout (FORKO) mouse. Biology of Reproduction 2000 62:1146-1159.

Weinbauer GF, Aslam H, Krishnamurthy H, Brinkworth MH, Einspanier A, Hodges JK. Quantitative analysis of spermatogenesis and apoptosis in the common marmoset (Callithrix jacchus) reveals high rates of spermatogonial turnover and high spermatogenic efficiency. Biology of Reproduction 2001 64:120-126.